The Significance of Faecal Indicators in Water
A Global Perspective

The Significance of Faecal Indicators in Water

A Global Perspective

Edited by

David Kay
CREH, Institute of Geography & Earth Sciences, Aberystwyth University, Aberystwyth, Ceredigion, U.K.
Email : dave@crehkay.demon.co.uk

Colin Fricker
CRF Consulting, Reading, Berkshire, U.K.

RSC Publishing

The proceedings of the conference Faecal Indicators: Problem or Solution? Has technical progress reduced the need for faecal indicators? held on 6th to 8th June 2011 at Edinburgh Conference Centre, Heriot Watt University, UK

Special Publication No. 337

ISBN: 978-1-84973-169-0

A catalogue record for this book is available from the British Library

Published by The Royal Society of Chemistry,
Thomas Graham House, Science Park, Milton Road,
Cambridge CB4 0WF, UK

Registered Charity Number 207890

For further information see our web site at www.rsc.org
Printed and bound in Great Britain by CPI Group (UK) Ltd, Croydon, CR0 4YY, UK

UNIVERSITY OF STRATHCLYDE
30125 00855975 8

ANDERSONIAN LIBRARY
WITHDRAWN
FROM
LIBRARY
STOCK
UNIVERSITY OF STRATHCLYDE

PREFACE

This book derives from a three day international meeting in Edinburgh, Scotland held in June 2011. It was the second such meeting organised by the Royal Society of Chemistry, the Society of Chemical Industry (SCI) and the Institute of Water with the first being held in 1995 in Leeds, England. Delegates from 28 countries attended this Edinburgh event.

The meeting format included submitted papers by the science and management communities as well as formally scheduled open discussion sessions designed to explore key contentious themes emerging in this dynamic and expanding area of science. The key themes addressed were:

- Coliforms and pathogens in drinking water
- Health risks from environmental water
- Emerging methods
- Sustainable water and wastewater systems
- The way forward and challenges we need to address

This book's 18 chapters present both review articles and research findings presented at the meeting as well as invited poster-based papers. Its readership will comprise water scientists, microbiologists, food scientists, regulators and managers.

The book charts a dynamic and rapidly evolving area of science in which the 'demise' of the microbial indicator has long been forecast as imminent in the face of rapidly developing methods for direct pathogen identification. In fact, its resilience as an operational and robust management tool has been remarkable and it is perhaps surprising to many involved in this series of meetings that much of the debate still centres on indicator concepts rather than direct enumeration of pathogen species as real-time risk assessment systems.

The meeting was sponsored by IDEXX, Scottish Water, Bio-Rad, Biomerieux, Oxoid and Dwr Cymru Welsh Water. Excellent administration of the whole project was provided by Maggi Churchouse Events and SCI.

The organising committee for the event were Clive Thompson, John Farmer, Colin Fricker, Bill Keevil, Melinda Maux, Gordon Nichols, Robert Pitchers, Keith Smith, Samantha Vince, Simon Gillespie, Andy Headland and David Kay.

David Kay
Colin Fricker

Contents

FAECAL INDICATORS AND PATHOGENS: EXPANDING OPPORTUNITIES FOR THE MICROBIOLOGY COMMUNITY

D. Kay[1], J. Crowther[2], C. Davies[1], A. Edwards[3], L. Fewtrell[1] , C. Francis[3], C. Kay[1], A. McDonald[4], C. Stapleton[1], J. Watkins[3], M. Wyer[1].

[1] CREH, Institute of Geography & Earth Sciences, Aberystwyth University, Aberystwyth, Ceredigion, SY23 3DB, U.K.
[2]CREH, University of Wales: Trinity St David, Lampeter, Ceredigion, SA48 7ED, U.K.
[3]CREH *Analytical* Ltd, Hoyland House, 50 Back Lane, Horsforth, Leeds LS18 4RF, U.K.
[4]School of Earth and Environment, University of Leeds, Leeds S2 9JT, U.K.

1 INTRODUCTION

Demands for a high quality science evidence-base and policy support by regulators, government and operational managers is increasing rapidly in the field of faecal indicators and pathogens. This dynamic is driven by emerging regulatory paradigms in North America, Europe and Austral-Asia underpinned by international science led organisations such as the WHO[1,2] (the World Health Organisation). Central to this development is the requirement to manage microbial risks in an integrated manner at the catchment scale. This implies quantification of diverse pollutant sources with very different flux quantity and timing. The complex spatial and temporal input pattern of microbial flux then undergoes complex processes causing attenuation, and possibly regrowth, in many catchment compartments such as the land surface, soil systems, groundwater, river waters and sediments and, thence, in estuarine and near-shore systems. Notwithstanding this complexity, which has received very little research attention when compared to: for example, the nutrient parameters, regulators in North America and Europe are required to develop a catchment scale Total Maximum Daily Load (TMDL-USA) estimate or a Programme of Measures (POM-EU) respectively to ensure resource use locations in rivers, lakes and/or near-shore waters comply with microbial standards. It is these legislative drivers: namely the United States Clean Water Act[3] (USCWA) and European Union Water Framework Directive[4] (EUWFD) which have produced immediate and increasing pressure for high quality science policy support and research within the science community focusing on environmental microbiology.

This area is central to managing water resources at the catchment scale. The best recent evidence of this pivotal position is seen in the summary data produced by USEPA which provide a real-time summary of the reasons for water quality impairments (non-compliance in EU terminology) (Figure 1) and the numbers of resultant TMDLs completed (Figure 2)[5,6]. Nearly all these impairments are due to non-compliance of recreational and shellfish growing waters with faecal coliform regulatory standards designed to protect public health. The US experience offers some 20 years of catchment-scale water quality regulation prior to the implementation of parallel European Union legislation in the form of the EUWFD. The emerging US evidence-base places microbial pollution at the fore-front of water quality concerns and similar prioritisation is likely in Europe where standards also derive largely from WHO Guidelines and publications covering drinking[2], recreational[7,8] and shellfish harvesting waters[9].

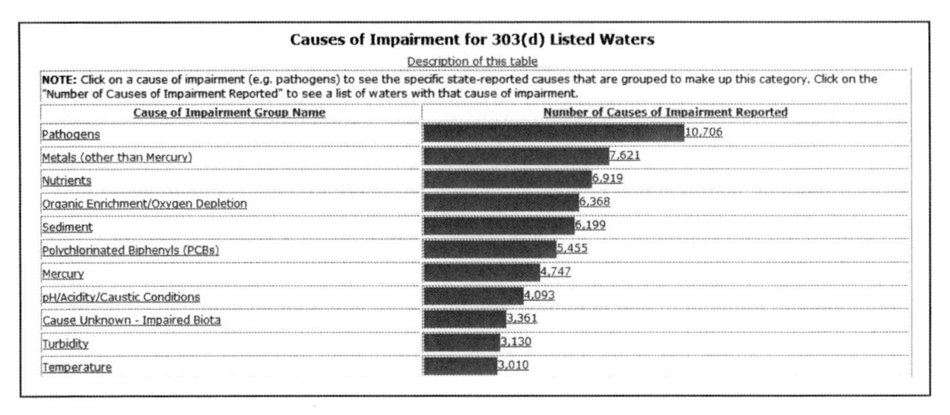

Causes of Impairment for 303(d) Listed Waters

Description of this table

NOTE: Click on a cause of impairment (e.g. pathogens) to see the specific state-reported causes that are grouped to make up this category. Click on the "Number of Causes of Impairment Reported" to see a list of waters with that cause of impairment.

Cause of Impairment Group Name	Number of Causes of Impairment Reported
Pathogens	10,706
Metals (other than Mercury)	7,621
Nutrients	6,919
Organic Enrichment/Oxygen Depletion	6,368
Sediment	6,199
Polychlorinated Biphenyls (PCBs)	5,455
Mercury	4,747
pH/Acidity/Caustic Conditions	4,093
Cause Unknown - Impaired Biota	3,361
Turbidity	3,130
Temperature	3,010

Figure 1 *Reasons for water quality impairments reported by USEPA in December 2011*

National Cumulative TMDLs by Pollutant

This chart includes TMDLs since October 1, 1995.

Description of this table

NOTE: Click on the underlined "Pollutant Group" value to see a detailed list of pollutants. Click on the underlined "Number of TMDLs" value to see a listing of those TMDLs for the pollutant Group.

Pollutant Group	Number of TMDLs	Number of Causes of Impairment Addressed
Pathogens	10,125	10,391
Metals (other than Mercury)	7,998	8,173
Mercury	6,970	7,002
Nutrients	5,278	6,285
Sediment	3,860	4,437
Organic Enrichment/Oxygen Depletion	1,943	2,051
Temperature	1,930	1,941

Figure 2 *Completed TMDL studies completed by USEPA in December 2011*

The emergence of this policy agenda is placing new challenges on managers of catchment activities from the farming community through to urban waste water treatment authorities. For both groups, microbial pollution is a growing concern. For the livestock farming community, the realisation that livestock contributions of faecal indicators to resource use sites is, first, highly episodic and, second, can exceed catchment-scale human sewage-derived fluxes has been challenging[10,11,12,13,14]. For the sewage undertakers, the uncomfortable realisation that their traditional suite of regulatory parameters: namely biochemical oxygen demand, suspended sediments and ammoniacal nitrogen; do not provide an indication of microbial flux and certainly do not ensure compliance with microbial standards despite being termed the 'sanitary parameters' by the engineering profession.

2 CATCHMENT-SCALE MICROBIAL FLUX

Kay et al.[15] have defined four principal sources of microbial loadings to rivers and coastal waters, namely:

1. human sewage disposal systems which discharge via pipes known as 'point-source' discharges. These can be further split into:
   a. 'continuous discharges' of sewage effluent following various levels of treatment from simple screening of solids (e.g. rags and plastics) to full biological treatment and disinfection. The receiving waters can include rivers, lakes or coastal waters through outfall pipes; and
   b. 'intermittent discharges' which occur when the sewerage system is receiving urban surface water drainage after rainfall. This can increases the flow beyond the capacity of the sewer, thus, producing short-term discharges of untreated, but diluted, raw sewage to rivers, lakes and coastal waters. This is normally discharged through:
      i. 'combined sewer overflows' in the sewer line itself,
      ii. 'storm tank overflows', generally sited just upstream of waste water treatment works, and
      iii. 'pumping station overflows', which operate when a pumping station on the sewer line is overwhelmed because of rainfall-supplemented flow volume. These overflows may also operate during dry weather conditions in response to emergency conditions such as pumping station failure or sewer blockages.
2. livestock-derived microbial fluxes deriving from a range of catchment sources commonly termed 'diffuse-source pollution'[10]. The pattern of this loading is site specific but includes:
   a. voiding of faeces to land (and, in some, cases to water) by livestock at;
      i. in-field feeding and drinking points;
      ii. gates between fields used for daily stock movements, notably for dairy farming operations where stock may be moved between milking facilities and fields, particularly in the summer period;
      iii. river crossing points which can be in the form of simple fords or bridges; and
      iv. stream bank drinking points, where faeces may be voided directly into the watercourse or onto land which has been highly trampled and 'puddled/poached' by livestock, creating a mobile and available faecal store.
   b. Faeces directly voided onto farm yards and other hardstanding areas which stock use during farming activities such as milking. Some of this loading will commonly be scraped and delivered to a dedicated manure

or slurry store (see 2c below) but residues remaining after normal yard cleaning will generally be washed from the impervious farm hardstanding area, often entering small drainage ditches and watercourses, particularly where the farm hardstandings have good hydrological connectivity to the catchment drainage network.

c. Livestock manure and slurry, which accumulates from housed livestock, mainly during the winter period in temperate climates, and is commonly stored in slurry tanks or lagoons. This is spread onto fields as a fertilizer and can be timed to maximise feed crop growth. However, sometimes the imperative to spread slurry is driven by the operational requirement to increase available storage capacity after periods of high rainfall, which can encourage inappropriate spreading at high risk periods. Where slurry is applied to fields with good hydrological connectivity to adjacent streams, the microbial loading to the stream environment can be very high, with associated problems of high biochemical oxygen demand from organic-rich material input and high nutrient loadings.

3. Wildlife populations in catchment systems are another potential source of microbial pathogen loading. Generally, microbial loadings from wildlife in intensively farmed catchments are low compared to the livestock and/or human population inputs. However, this area has not received intensive research attention[16] and some surprising findings have been reported, suggesting that roof runoff, commonly perceived to be a clean and sustainable resource, can be highly contaminated with faecal indicator organisms (FIOs), such as intestinal enterococci from avian wildlife sources[17].
4. Urban diffuse microbial pollution presents a further source. This comprises street drainage and associated drain flow, generally separate from the foul sewerage system, which may be contaminated with faecal matter deposited onto urban roads, pavements and roofs. Canine and feline pets certainly contribute to this loading, as do avian and rodent urban wildlife populations. This is a difficult loading to quantify because many urban areas may also have a number of inappropriate and 'informal' connections of foul sewage from individual properties into the non-foul surface water drainage systems. Links between the foul sewer and surface water drainage networks may also result from ageing and poorly maintained infrastructure. Thus, quantification of the true quality of urban diffuse microbial fluxes is problematical because identification of sites which can be guaranteed free of any such mis-connections in the urban upstream infrastructure is extremely difficult.

Operational management questions required to disentangle this complex web of catchment flux and deliver health risk management through regulation of microbial parameters centres on a series of current key science agendas and questions, namely:

i. Can we quantify microbial flux given the complex catchment source patterns evident and the extreme temporal variability characteristic of microbial flux which is often driven by rainfall events and/or infrastructure breakdown?
ii. Can catchment-scale modelling tools be developed to inform (i) and (iii) and also guide complex management decisions needed to deliver cost effective attenuation of microbial flux to improve water quality compliance at the

catchment scale, often involving difficult trade-offs between different users and interests?

iii. Can we determine the source species of faecal indicators at the measurement, or compliance, location using source tracking tools now available or under development and can this approach quantify the relative contribution of, for example, human and ruminant microbial contributions?

iv. Can near-real-time analytical protocols be developed to shorten the times associated with microbial culture methods and, thus, deliver operationally appropriate, and timely, information to managers seeking to limit and control health risk?

v. Can real-time prediction of microbial pollution episodes replace and/or supplement traditional sampling of faecal indicators at resource use sites?

This chapter presents evidence for UK empirical studies relevant to questions (i) and (ii). Questions (iii), (iv) and (v) are addressed subsequent chapters.

3 QUANTITATIVE MICROBIAL SOURCE APPORTIONMENT

A number of recent studies have reported catchment-scale faecal indicator (coliforms and/or enterococci) flux and sought to quantify multiple source contributions. In the United Kingdom, compliance with revised criteria for bathing waters published by the EU, which are based on WHO health-based guidelines, has driven a series of such investigations identified in Figure 3.

These studies have often focused on a key period, such as the bathing season, and the key challenge has been to design a field data acquisition protocol able to characterise the very short term changes in faecal indicator concentrations caused by episodic changes in concentration, caused by rainfall-driven events. This is challenging, particularly where aseptic sample collection is compromised by the use of automated sampling systems which are often susceptible to carry-over problems between discrete sampling events. The approach adopted in these UK studies has been station a field team on site for the project duration to ensure aseptic 'hand' collection of opportunistic event-based samples during high flow events as well as routinely planned base flow samples at all catchment outlets and at key infra-structure locations. Together with flow measurement of streams and key infrastructure sites this allows construction of flux pie charts (Figure 4) and temporal contribution plots (Figure 5) which illustrate the relative contribution of point and diffuse source fluxes on an hourly basis through the study periods or bathing season. Presentation tools as illustrated in Figure 5 provide key management information describing the flux contributions during the principal compliance stress periods: i.e. during episodic rainfall events which are a 'normal' weather-driven characteristic of many bathing waters and shellfish harvesting sites world-wide.

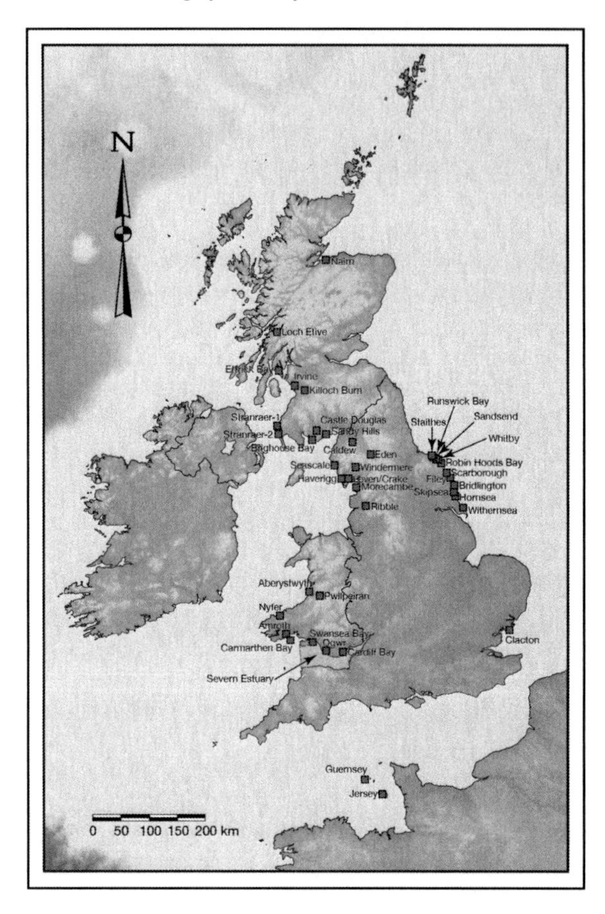

Figure 3 *Recent UK catchment-scale quantitative microbial source apportionment studies driven by bathing and shellfish water compliance.*

Studies such as the Etive investigation[18] are complex and expensive to deliver. This has led to attempts to provide simple modelling strategies designed to use remote sensing tools to predict faecal indicator flux from infrastructure and diffuse sources. The principal infrastructure 'predictors' are the treatment type and population equivalent served by the treatment plants within the catchment and the drivers of diffuse pollution are the land use and land area under each land use. Published data on the microbiological quality of sewage effluents under different treatment regimes (Table 1) and catchment export coefficients for faecal indicators for common UK land uses types provide the base data for this type of desk-study which has been used to derive estimates of agricultural microbial source apportionment for key UK catchments draining to the coast (Figure 6).

Although empirically-based, these estimates remain approximations when applied to areas where no dedicated sampling programme has been undertaken. The experience of the studies identified in Figure 3 strongly suggests that intermittent flux from the sewerage infrastructure in the form of combined sewage overflows and storm tank overflows is a remaining area of considerable uncertainty, at least in the UK. These are rarely measured and recorded in real-time and quantitative estimates of flux from such discharges are often dependent on sewer modelling investigations which may not provide accurate estimates of flux timing and volume derived from specific overflow events.

These basic 'black-box' modelling tools have some operational utility but do not easily facilitate scenario modelling to determine the likely implications of alternative management intervention strategies. Ideally, this requires a more process-based, or 'white-box', modelling approach. However, catchment microbial science lags well behind modelling of other parameters such as dissolved oxygen, sediments and the nutrient parameters and the parameterisation of white box microbial models is in its infancy as a science agenda[13]. Thus, empirically based parameterisation data and relationships are not available credibly to populate a process-based modelling approach at this time although progress is accelerating in this area and significant advances are underway world-wide.

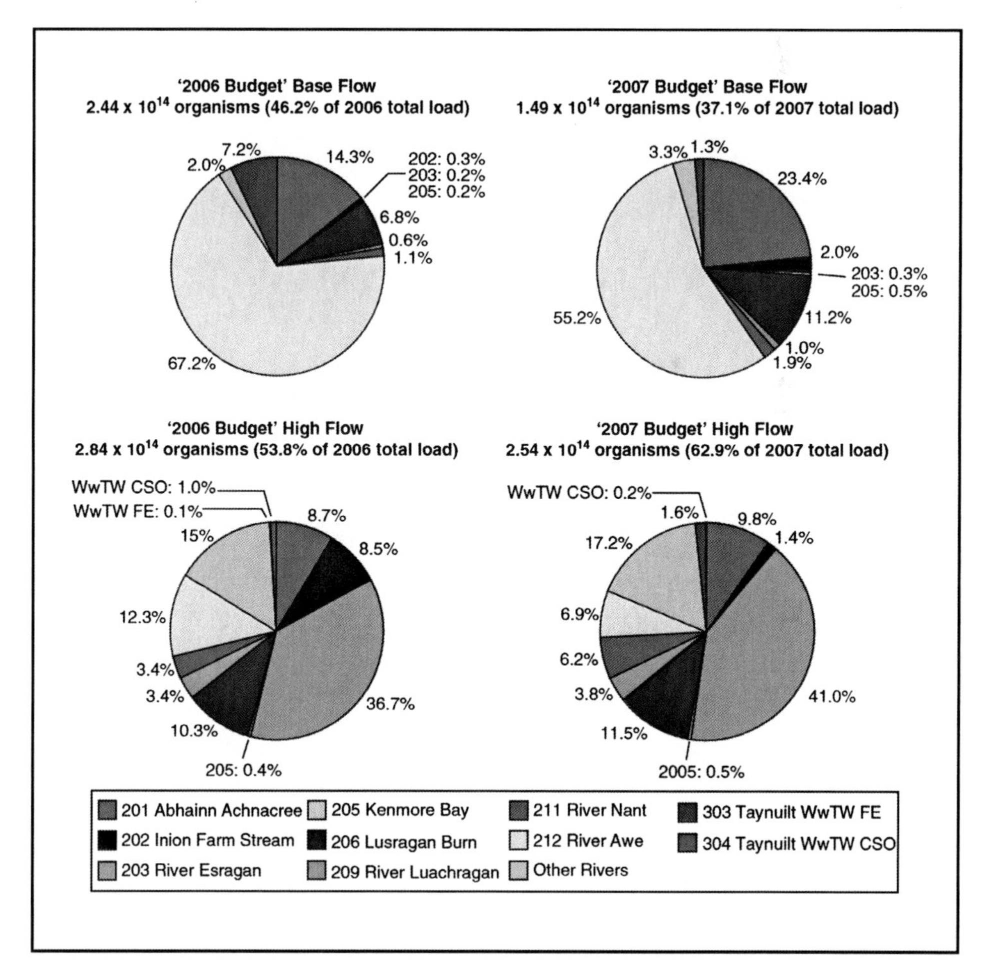

Figure 4 *Base flow and high flow faecal coliform budgets for a two year field sampling programme of multiple sources discharging to Loch Etive, Scotland (see Figure 2).*

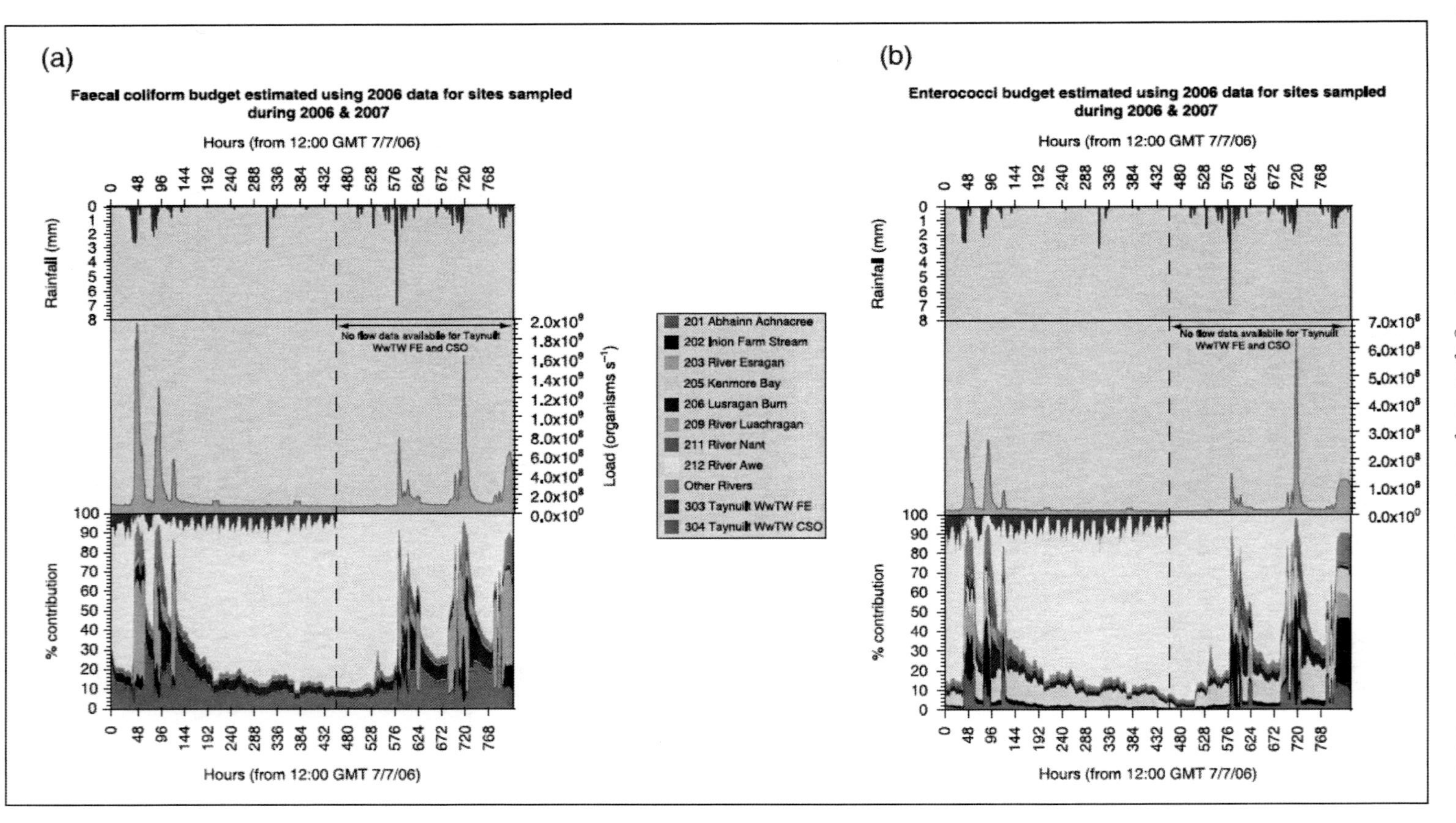

Figure 5 *Hourly rainfall (mm) at Inion Farm in the Loch Etive Study showing instantaneous load (organisms per second) and proportional contribution (%) of organisms to the hourly load input for: (a) faecal coliforms; (b) enterococci*[18].

Table 1 *Summary of faecal indicator organism concentrations (cfu 100 ml^{-1}) for different sewage treatment levels and individual types of sewage-related effluents under different flow conditions: geometric means (GMs), 95% confidence intervals (CIs)[a]; and results of t-tests comparing base- and high-flow GMs for each group and type[b]; and (in footnote) results of t-tests comparing GMs for the two untreated discharge types and the two tertiary-treated effluent types[19].*

Indicator organism Treatment levels and specific types	Base flow conditions: n^c	Geometric mean	Lower 95% CI	Upper 95% CI	High flow conditions: n^c	Geometric mean	Lower 95% CI	Upper 95% CI
TOTAL COLIFORMS								
Untreated	**253**	**3.9×10^7 *(+)**	**3.2×10^7**	**4.6×10^7**	**279**	**8.2×10^6 *(–)**	**7.0×10^6**	**9.6×10^6**
Crude sewage discharges[d]	253	3.9×10^7 *(+)	3.2×10^7	4.6×10^7	79	1.2×10^7 *(–)	8.2×10^6	1.6×10^7
Storm sewage overflows[d]					200	7.2×10^6	5.9×10^6	8.4×10^6
Primary	**130**	**3.0×10^7 *(+)**	**2.3×10^7**	**3.9×10^7**	**14**	**1.2×10^7 *(–)**	**4.0×10^6**	**3.7×10^7**
Primary settled sewage	61	3.8×10^7	3.0×10^7	4.7×10^7	8	2.2×10^7	–	–
Stored settled sewage	26	2.4×10^7	1.2×10^7	5.1×10^7	1	1.1×10^6	–	–
Settled septic tank	43	2.5×10^7	1.3×10^7	4.2×10^7	5	7.5×10^6	–	–
Secondary	**853**	**1.1×10^6**	**9.5×10^5**	**1.2×10^6**	**183**	**1.3×10^6**	**1.0×10^6**	**1.7×10^6**
Trickling filter	472	1.4×10^6	1.2×10^6	1.7×10^6	76	1.4×10^6	1.0×10^6	1.9×10^6
Activated sludge	256	7.8×10^5 *(–)	6.2×10^5	1.0×10^6	92	1.4×10^6 *(+)	8.6×10^5	2.1×10^6
Oxidation ditch	35	8.1×10^5	4.6×10^5	1.4×10^6	5	3.1×10^6	–	–
Trickling/sand filter	10	6.4×10^5	2.8×10^5	1.4×10^6	8	2.7×10^5	–	–
Rotating biological contactor	80	6.8×10^5	4.6×10^5	1.0×10^6	2	4.0×10^6	–	–
Tertiary	**182**	**5.5×10^3**	**3.4×10^3**	**9.0×10^3**	**8**	**3.8×10^3**	–	–
Reedbed/grass plot[e]	73	3.7×10^4	1.5×10^4	8.1×10^4	2	2.3×10^4	–	–
Ultraviolet disinfection[e]	109	1.5×10^3	9.9×10^2	2.6×10^3	6	2.1×10^3	–	–
FAECAL COLIFORMS								
Untreated	**252**	**1.7×10^7 *(+)**	**1.4×10^7**	**2.0×10^7**	**282**	**2.8×10^6 *(–)**	**2.3×10^6**	**3.2×10^6**
Crude sewage discharges[d]	252	1.7×10^7 *(+)	1.4×10^7	2.0×10^7	79	3.5×10^6 *(–)	2.6×10^6	4.7×10^6
Storm sewage overflows[d]					203	2.5×10^6	2.0×10^6	2.9×10^6
Primary	**127**	**1.0×10^7 *(+)**	**8.4×10^6**	**1.3×10^7**	**14**	**4.6×10^6 *(–)**	**2.1×10^6**	**1.0×10^7**
Primary settled sewage	60	1.8×10^7	1.4×10^7	2.1×10^7	8	5.7×10^6	–	–
Stored settled sewage	25	5.6×10^6	3.2×10^6	9.7×10^6	1	8.0×10^5	–	–
Settled septic tank	42	7.2×10^6	4.4×10^6	1.1×10^7	5	4.8×10^6	–	–

Indicator organism Treatment levels and specific types	Base flow conditions: n^c	Geometric mean	Lower 95% CI	Upper 95% CI	High flow conditions n^c	Geometric mean	Lower 95% CI	Upper 95% CI
Secondary	**864**	**3.3x10^5 *(–)**	**2.9x10^5**	**3.7x10^5**	**184**	**5.0x10^5 *(+)**	**3.7x10^5**	**6.8x10^5**
Trickling filter	477	4.3x10^5	3.6x10^5	5.0x10^5	76	5.5x10^5	3.8x10^5	8.0x10^5
Activated sludge	261	2.8x10^5 *(–)	2.2x10^5	3.5x10^5	93	5.1x10^5 *(+)	3.1x10^5	8.5x10^5
Oxidation ditch	35	2.0x10^5	1.1x10^5	3.7x10^5	5	5.6x10^5	–	–
Trickling/sand filter	11	2.1x10^5	9.0x10^4	6.0x10^5	8	1.3x10^5	–	–
Rotating biological contactor	80	1.6x10^5	1.1x10^5	2.3x10^5	2	6.7x10^5	–	–
Tertiary	**179**	**1.3x10^3**	**7.5x10^2**	**2.2x10^3**	**8**	**9.1x10^2**	**–**	**–**
Reedbed/grass plot[e]	71	1.3x10^4	5.4x10^3	3.4x10^4	2	1.5x10^4	–	–
Ultraviolet disinfection[e]	108	2.8x10^2	1.7x10^2	4.4x10^2	6	3.6x10^2	–	–
ENTEROCOCCI								
Untreated	**254**	**1.9x10^6 *(+)**	**1.6x10^6**	**2.3x10^6**	**280**	**4.9x10^5 *(–)**	**4.2x10^5**	**5.6x10^5**
Crude sewage discharges[d]	254	1.9x10^6 *(+)	1.6x10^6	2.3x10^6	79	8.9x10^5 *(–)	6.7x10^5	1.2x10^6
Storm sewage overflows[d]					201	3.8x10^5	3.2x10^5	4.5x10^5
Primary	**128**	**1.3x10^6**	**1.1x10^6**	**1.7x10^6**	**14**	**9.8x10^5**	**4.4x10^5**	**2.2x10^6**
Primary settled sewage	61	2.4x10^6	2.1x10^6	2.7x10^6	8	1.9x10^6	–	–
Stored settled sewage	26	6.2x10^5	3.2x10^5	1.1x10^6	1	2.9x10^5	–	–
Settled septic tank	41	9.3x10^5	5.3x10^5	1.6x10^6	5	4.3x10^5	–	–
Secondary	**871**	**2.8x10^4 *(–)**	**2.5x10^4**	**3.2x10^4**	**182**	**4.7x10^4 *(+)**	**3.6x10^4**	**6.1x10^4**
Trickling filter	483	4.1x10^4	3.5x10^4	4.7x10^4	76	5.7x10^4	4.2x10^4	8.3x10^4
Activated sludge	262	2.1x10^4 *(–)	1.8x10^4	2.7x10^4	91	4.1x10^4 *(+)	2.7x10^4	6.0x10^4
Oxidation ditch	35	2.0x10^4	1.0x10^4	4.0x10^4	5	1.2x10^5	–	–
Trickling/sand filter	11	2.1x10^4	1.0x10^4	5.3x10^4	8	1.1x10^4	–	–
Rotating biological contactor	80	9.6x10^3	6.7x10^3	1.4x10^4	2	3.7x10^5	–	–
Tertiary	**177**	**3.0x10^2**	**1.8x10^2**	**5.0x10^2**	**8**	**2.1x10^2**	**–**	**–**
Reedbed/grass plot[e]	73	1.9x10^3	7.1x10^2	4.3x10^3	2	2.3x10^3	–	–
Ultraviolet disinfection[e]	104	8.3x10^1	4.6x10^1	1.1x10^2	6	9.7x10^1	–	–

[a]CIs only reported where $n \geq 10$

[b]*t*-tests comparing low- and high-flow GM concentrations only undertaken where $n \geq 10$ for both sets of samples; only statistically significant ($p < 0.05$) differences between base- and high-flow GM concentrations are reported: indicated by *, with the higher GM being identified as *(+) and the lower value by *(–)

[c]*n* indicates number of valid enumerations, which in some cases may be less than the actual number of samples

[d]*t*-tests comparing the GM concentrations between the two untreated discharge types show high-flow GM concentrations to be significantly higher in crude sewage discharges than storm sewage overflows for TC ($p < 0.05$) and EN ($p < 0.001$)

[e]*t*-tests comparing the GM concentrations between the two tertiary-treatment effluent types show GM TC, FC and EN concentrations to be significantly higher ($p < 0.001$) in reedbed/grass plot effluents than effluents from UV disinfection for base-flow conditions (there are too few high-flow samples for these tertiary effluents for meaningful comparisons to be made for high-flow GM concentrations)

Table 2 *Summary of geometric mean faecal indicator organism (FIO) export coefficients (log10 cfu km-2 hr-1) under base- and high-flow conditions at 205 UK sub-catchment sampling points and for various subsets, and results of paired, 1-tailed t-tests to establish whether there are significant elevations at high flow compared with base flow20.*

FIO **Sub-catchment land use**	***n***	**Base flow: Geometric mean**	**Lower 95% CI**	**Upper 95% CI**	**High flow: Geometric mean**[a]	**Lower 95% CI**	**Upper 95% CI**
TOTAL COLIFORMS							
All subcatchments	205	$1.8x10^9$	$1.4x10^9$	$2.4x10^9$	$9.5x10^{10}$**	$7.2x10^{10}$	$1.2x10^{11}$
Degree of urbanisation[b]							
Urban	20	$8.5x10^9$	$3.3x10^9$	$2.2x10^{10}$	$4.1x10^{11}$**	$1.6x10^{11}$	$1.1x10^{12}$
Semi-urban	60	$4.2x10^9$	$2.6x10^9$	$6.7x10^9$	$1.5x10^{11}$**	$8.3x10^{10}$	$2.7x10^{11}$
Rural	125	$9.3x10^8$	$6.9x10^8$	$1.3x10^9$	$6.1x10^{10}$**	$4.6x10^{10}$	$8.0x10^{10}$
Rural subcatchments with different dominant land uses							
≥ 75% Improved pasture	15	$2.9x10^9$	$1.4x10^9$	$6.0x10^9$	$2.8x10^{11}$**	$1.6x10^{11}$	$4.9x10^{11}$
≥ 75% Rough grazing	13	$7.1x10^8$	$3.5x10^8$	$1.4x10^9$	$5.3x10^{10}$**	$2.6x10^{10}$	$1.1x10^{11}$
≥ 75% Woodland	6	$3.1x10^8$	$5.7x10^7$	$1.6x10^9$	$1.4x10^{10}$**	$6.0x10^9$	$3.4x10^{10}$
FAECAL COLIFORMS							
All subcatchments	205	$5.5x10^8$	$4.1x10^8$	$7.2x10^8$	$3.6x10^{10}$**	$2.7x10^{10}$	$4.8x10^{10}$
Degree of urbanisation[b]							
Urban	20	$2.8x10^9$	$1.1x10^9$	$7.2x10^9$	$1.3x10^{11}$**	$4.8x10^{10}$	$3.6x10^{11}$
Semi-urban	60	$1.2x10^9$	$7.4x10^8$	$1.9x10^9$	$4.6x10^{10}$**	$2.5x10^{10}$	$8.6x10^{10}$
Rural (< 2.5% built-up land)	125	$2.9x10^8$	$2.1x10^8$	$4.0x10^8$	$2.6x10^{10}$**	$1.9x10^{10}$	$3.5x10^{10}$
Rural subcatchments with different dominant land uses							
≥ 75% Improved pasture	15	$8.3x10^8$	$4.3x10^8$	$1.6x10^9$	$1.2x10^{11}$**	$6.5x10^{10}$	$2.2x10^{11}$
≥ 75% Rough grazing	13	$2.5x10^8$	$1.1x10^8$	$5.7x10^8$	$2.5x10^{10}$**	$1.1x10^{10}$	$5.5x10^{10}$
≥ 75% Woodland	6	$2.0x10^7$	$4.7x10^6$	$8.2x10^7$	$3.3x10^9$**	$1.3x10^9$	$8.8x10^9$

FIO **Sub-catchment land use**	*n*	**Base flow: Geometric mean**	**Lower 95% CI**	**Upper 95% CI**	**High flow: Geometric mean**[a]	**Lower 95% CI**	**Upper 95% CI**
ENTEROCOCCI							
All subcatchments	205	8.3×10^{7}	6.6×10^{7}	1.1×10^{8}	7.1×10^{9}**	5.5×10^{9}	9.3×10^{9}
Degree of urbanisation[b]							
Urban	20	4.0×10^{8}	2.1×10^{8}	7.6×10^{8}	2.7×10^{10}**	1.1×10^{10}	6.2×10^{10}
Semi-urban	60	1.5×10^{8}	9.8×10^{7}	2.2×10^{8}	1.1×10^{10}**	6.1×10^{9}	1.9×10^{10}
Rural (<2.5% built-up land)	125	4.9×10^{7}	3.7×10^{7}	6.5×10^{7}	4.7×10^{9}**	3.5×10^{9}	6.3×10^{9}
Rural subcatchments with different dominant land uses							
≥ 75% Improved pasture	15	9.6×10^{7}	5.2×10^{7}	1.8×10^{8}	2.2×10^{10}**	1.3×10^{10}	3.8×10^{10}
≥ 75% Rough grazing	13	3.3×10^{7}	1.2×10^{7}	9.0×10^{7}	3.6×10^{9}**	1.3×10^{9}	9.7×10^{9}
≥ 75% Woodland	6	8.5×10^{6}	3.8×10^{6}	1.9×10^{7}	3.8×10^{8}**	1.3×10^{8}	1.1×10^{9}

[a]Significant elevations in export coefficients at high flow are indicated: ** $p < 0.001$
[b]Degree of urbanisation, categorised according to percentage built-up land: 'Urban' (≥ 10.0%), 'Semi-urban' (2.5-9.9%) and 'Rural' (<2.5%)

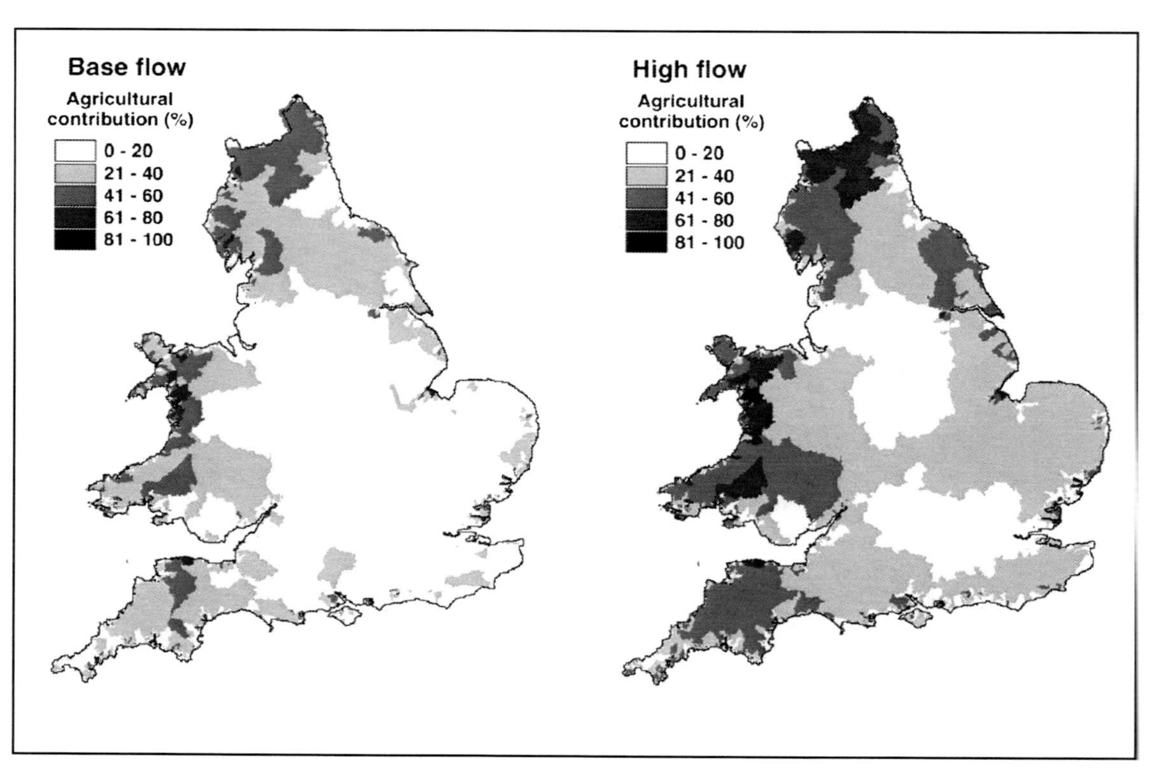

Figure 6 *Predicted percentages of GM faecal coliform concentrations in rivers draining to the coast in England and Wales derived from agricultural sources, compared with sewerage-related 'urban' sources under base- and high-flow conditions during the summer bathing season*[21].

4 MICROBIAL FLUX ATTENUATION AND CONTROL

Notwithstanding the lack of fully parameterised, and process-based, modelling systems, the regulatory and operational communities are faced with the challenge of achieving compliance at water use sites against microbial standards using the coliform and enterococci indicator bacteria. Many studies world-wide have addressed this problem with catchment-scale and plot scale investigations which were recently reviewed by Kay et al. They considered six broad categories of Best Management Practice (BMPs), or intervention used to reduce microbial flux to catchment outlets, namely: ponds; vegetated treatment areas; integrated constructed wetlands; woodchip corrals; vegetated riparian buffer strips; and finally in-stream ponds. Figure 7 presents the median attenuations observed and the ranges reported in the literature available.

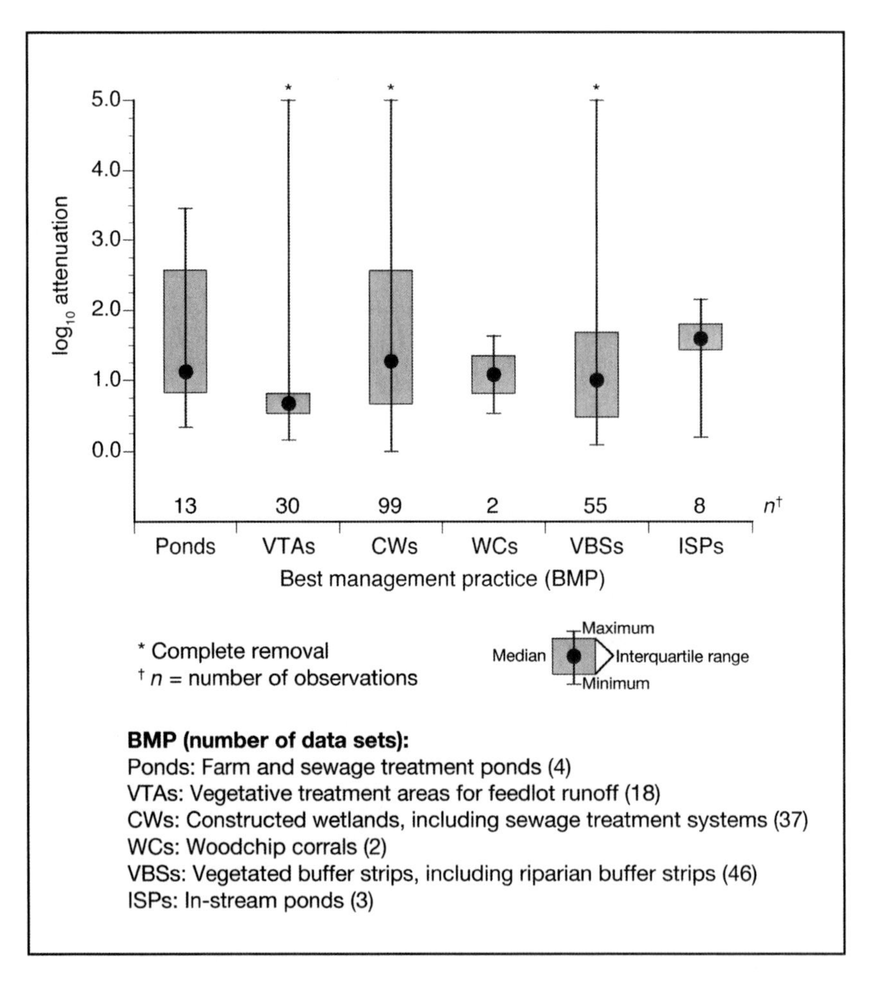

Figure 7 *Summary FIO attenuation data for the six BMPs for which datasets have been compiled (study numbers in parenthesis)*[22].

The principal feature of all BMPs examined to date is the extreme range in $\log_{10}$ reductions in faecal indicator concentration observed across the installed BMPs. This makes clear policy guidance on selection design of the most effective suite of control measure very difficult at this time. To some extent, this reflects the sourcing of this empirical data resource and the environmental range of sites examined. However, it reinforces the need for detailed empirical data acquisition on BMP effectiveness in the regions where attenuation design strategies are required.

5 CONCLUSIONS

Catchment microbial dynamics is an emerging new science agenda being defined by clear policy-led demand for a credible evidence-base to inform appropriate remediation of pollution causing impairment, or non-compliance, of water use sites world-wide. Resolution of this microbial problem is becoming central to the catchment management communities in North America and Europe. Many of the issues raised by questions (i) to (v) above are address in contributions to this book and the conference from which it derives.

Acknowledgements

We are grateful to The Scottish Aquaculture Research Foundation, SEPA and Scottish Water who funded the study of Loch Etive and to Department for Environment Food and Rural Affairs DEFRA) who funded the work leading to the analyses presented in Figures 6 and 7. Tables 1 and 2 derive from the studies identified in Figure 1 and were funded by the European Commission (Interreg programmes), the Environment Agency for England and Wales, the Scottish Environmental Protection Agency (SEPA), the Scottish Government, United Kingdom Water Industry Research limited and many individual water companies. We are grateful to the many field sampling teams who have acquired water quality data in atrocious weather and, all seasons, to underpin this analysis.

References

1 WHO (2003) *Guidelines for safe recreational water environments Volume 1: Coastal and freshwaters* World Health Organisation, Geneva.

2 WHO (2011) *Guidelines for drinking water quality, Volume 2- Health criteria and other supporting information, 4th edn., World Health Organisation (WHO)* World Health Organisation, Geneva.

3 United Stated Government (1972) FEDERAL WATER POLLUTION CONTROL ACT, (33 U.S.C. 1251 et seq.) Ss Amended Through P.L. 107–303, November 27, 2002: available

4 Anon (2000) Council of the European Communities. Directive 2000/60/EC of the European Parliament and of the Council of 23 October 2000 establishing a framework for Community action in the field of water policy. *Official Journal of the European Union* L327, 1-72.

5 USEPA, U.E.P.A., 2005. Total Maximum daily loads US Environmental Protection Agency. USEPA, Washington DC. URL http://oaspub.epa.gov/waters/national_rept.control

6 Kay, D., Stapleton, C. M., Wyer, M. D., McDonald, A. T., and Crowther, J. (2006) Total Maximum Daily Loads (TMDL). The USEPA approach to managing faecal indicator fluxes to receiving waters: Lessons for UK environmental regulation? L. Gairns, C. Crighton, and W.A. Jeffrey (Eds) Agriculture and the Environment VI; Managing Rural Diffuse Pollution. Proceedings of the SAC/SEPA Biennial Conference, Edinburgh. International Water Association, Scottish Agricultural College, Scottish Environmental Protection Agency, Edinburgh, ISBN: 1-901322-63-7, pp. 23-33.

7 Anon (1976) Council of the European Communities. Council Directive 76/160/EEC of 8 December 1975 concerning the quality of bathing water. *Official Journal of the European Communities* L31, 1-7.

8 Anon (2006) Directive 2006/7/EC of The European Parliament and of The Council of 15th February 2006 concerning the management of bathing water quality and repealing Directive 76/160/EEC. *Official Journal of the European Union* L64 (4.3.2006), 37-51.

9 Rees, G., Pond, K., Kay, D. and Domingo, S. (2009) (Eds.) Safe *management of shellfish and harvest waters*. Published jointly by International Water Association and World Health Organization 358p. ISBN9781843392255.

10 Bos, R. and Bartram, J. (Eds) (2012) *Animal Waste, Water Quality and Human Health:* WHO - Emerging Issues in Water and Infectious Disease series. International Water Association and WHO, London. (*in press*).

11 Kay, D., Edwards, A.C., Ferrier, B., Francis, C., Kay, C., Rushby, L. Watkins, J., McDonald, A.T., Wyer, M., Crowther, J. and Wilkinson, J. (2007a) Catchment microbial dynamics: the emergence of a research agenda. *Progress in Physical Geography 31(1),* 59-76. (DOI:10.1177/0309133307073882).

12 Kay, D., McDonald, A.T., Stapleton, C.M., Wyer, M.D. and Crowther, J. (2009a) *Catchment to coastal systems: managing microbial pollutants in bathing and shellfish harvesting waters*. In Jenkins, A. and Ferrier, B. (Eds.) (2009) *Catchment Management Handbook.* Wiley-Blackwell, Chichester, ISBN 9781405171229, pp.181-208.

13 Ferguson, C. and Kay, D. (2011) *Transport of microbial pollution in catchment systems.* In Bos, R. and Bartram, J. (Eds) *Animal Waste, Water Quality and Human Health:* WHO - Emerging Issues in Water and Infectious Disease series. International Water Association and WHO, London. (*in press*).

14 Ashbolt, N., and Roser, D. (2003) Interpretation and management implications of event and baseflow pathogen data. In M. J. Pfeffer, D. Abs, and K. N. Brooks (Eds): *Watershed Management for Water Supply Systems*, American Water Resources Association, New York City, USA.

15 Kay, D., Crowther, J., Davies, C., Edwards, A., Fewtrell, L., Francis, C., Kay, C., McDonald, A., Stapleton, C. and Wyer, M. (2012) *Impacts of agriculture on water-borne pathogens.* In Harrison, R. (Ed) *Environmental impacts of modern agriculture.* Royal Society of Chemistry. (in press).

16 McDonald, A.T., Chapman, P.J. and Fukasawa, K. (2008) The microbial status of waters in a protected wilderness area. *Journal of Environmental Management 87*, 600-608

17 Edwards, A.C. and Kay, D. (2008) Farmyards an overlooked source of highly contaminated runoff. *Journal of Environmental Management 87*, 551-559. (doi:10.1016/j.jenvman.2006.06.027).

18 Stapleton, C.M., Kay, D., Wyer, M. D., Davies, C., Watkins, J., Kay, C., McDonald, A.T. and Crowther, J. (2010) Identification of microbial risk factors in shellfish harvesting waters: the Loch Etive case study. *Aquaculture Research* (Special Issue on the Proceedings of the Scottish Aquaculture: A Sustainable Future? Symposium, 21-22 April 2009, Edinburgh; ISSN *1*365-2109 doi:10.1111/j.1365-2109.2010.02666.x).

19 Kay, D., Stapleton, C.M., Crowther, J., Wyer, M.D., Fewtrell, L., Edwards, A., McDonald, A.T., Watkins, J., Francis, F. and Wilkinson, J. (2008b) Faecal indicator organism compliance parameter concentrations in sewage and treated effluents. *Water Research 42*, 442-454. (DOI:10.1016/j.watres.2007.07.036).

20 Kay, D., Crowther, J., Stapleton, C.M., Wyer, M.D., Anthony, S., Bradford, M., Edwards, A., Fewtrell, L., Francis, C., Hopkins, M., Kay, C., McDonald, A.T., Watkins, J. and Wilkinson, J. (2008c) Faecal indicator organism concentrations and catchment export coefficients in the UK. *Water Research 42*, 2649-2661. (DOI:10.1016/j.watres.2008.01.017).

21 Kay, D. Anthony, S., Crowther, J. Chambers, B., Nicholson, F., Chadwick, D., Stapleton, C. and Wyer, M. (2010a) Microbial water pollution: a screening tool for initial catchment-scale assessment and source apportionment *Science of the Total Environment 408,* 5649-5656. (Doi:10.1016/j.scitotenv.2009.07.033).

22 CREH (2010) *Source Strengths and Attenuation of Faecal indicators Derived from Livestock Farming Activities: Literature Review* Report on DEFRA Project WQ0203 Demonstration Test Catchments Initiative. 33 pp. http://randd.defra.gov.uk/Default.aspx?Menu=Menu&Module=More&Location=None&Completed=0&ProjectID=16470.

FAECAL INDICATORS IN DRINKING WATER – IS IT TIME TO MOVE ON?

Margaret McGuinness

Scottish Water, Castle House, Dunfermline, UK

1 INTRODUCTION

Assurance that drinking water is microbially safe has traditionally been determined by measuring bacterial indicators of water quality, most commonly total coliforms and *E. coli*. This chapter addresses four questions.

i. Should we continue to rely on these?
ii. Are these indicators sufficient to ensure microbial water quality?
iii. Should we adopt a more holistic approach?
iv. We have our risk management systems, our drinking water safety plans; does this mean that we could comfortably move, by focussing more on risk management, away from our reliance on end point testing?

Total coliforms have been shown to be a poor parameter for measuring the potential for faecal contamination of drinking water due to their presence as natural inhabitants of soil and water environments, their ability to grow in distribution systems and their inconsistent presence in water supplies during outbreaks of waterborne disease. All these points make it difficult to interpret the sanitary significance of their presence (in the absence of *E. coli*) or have confidence in water quality in their absence.

The presence of *E. coli* in drinking water is still considered to indicate faecal contamination. *E. coli* monitoring as a verification of water quality is a useful tool within a risk management system approach to water quality. There are, however, other indicators, both microbial and physical, we can use to monitor both drinking water system operation and performance and which provide better support to system management.

The question is – could we remove total coliforms as an indicator organism and use *E. coli* as the primary indicator of faecal contamination?

2 BACKGROUND

Snow proved water to be a vehicle for disease in 1855, during the late 1800's early 1900's knowledge about the nature and cause of disease increased rapidly and there was major focus on public health reform. As a consequence, techniques to identify and enumerate causative agents of disease were developed.
For more than 100 years, the microbial safety of drinking water has primarily been determined by testing for bacterial indicators of faecal pollution, mainly *E. coli* and total

coliforms. We use these to assess the potential public health risk of drinking water and their presence or absence are the key elements of drinking water guidelines.

Why do we test for total coliforms? The rationale presented is that this functional group of bacteria is present in large numbers in the gut of humans and other warm blooded animals. Hence, if water is contaminated by faeces, coliforms can be detected even after extensive dilution. However, this group lacks specificity. Of the total coliform group *E. coli* is the most numerous in mammalian faeces and is considered to represent the presence of faecal contamination; the presence of this organism also indicates that pathogenic bacteria, viruses and protozoa may be present. The drawback on relying on *E. coli* is that it is a poor indicator of the presence of viruses and parasitic protozoa that can survive for much longer periods than the indicator species.

A more holistic approach would be to increase attention on non-bacterial pathogens including *Cryptosporidium* spp. and viruses, effective testing of microbial water quality clearly requires more. We now look more closely at the turbidity of filtered waters and Ct values are increasingly used as indicators of quality and confidence can be further underpinned by the Drinking Water Safety Plan approach.

3 MICROBIAL INDICATORS

The qualities of a good indicator are.

- That they are universally present in the faeces of humans and warm blooded animals in large numbers.
- That they are readily detected by simple methods.
- That they do not grow in natural waters, the general environment or water distribution systems.
- That they persist in water and the extent to which they are removed by water treatment is similar to those of waterborne pathogens.

(WHO 1996)

With few exceptions, coliforms themselves are not considered to be a health risk, but their presence indicates that faecal pollution may have occurred and pathogens may be present as a result.

Total coliforms were originally used as a surrogate for *E. coli* for 3 reasons.

- Coliform bacteria were readily isolated from human faecal material and water that had been impacted by pollution.
- Most of the coliforms recovered from human faeces are *E. coli* and it was assumed that the presence of total coliforms reflected the presence of *E. coli*.
- The technology was not available easily to distinguish via routine analysis, *E. coli* from other coliforms in the early 1900's.

Hence, total coliforms were utilised and considered to be equivalent to *E. coli*, until more specific and rapid methods became available.
It was not until 1948 that the more specific and well known 48 hour test for faecal coliforms was accepted. Despite the development of this and other specific methods, the use of total coliforms was so common place that they were not dropped in favour of *E. coli* but have remained co-indicators.

As microbial understanding about the nature of disease and pathogens associated with disease has increased, techniques have been developed to isolate and enumerate pathogenic viruses and protozoa from water. However, these techniques are not specific, reliable or reproducible enough to give a reliable and inexpensive indication of water quality and so we still have a need for an indicator organism for routine use.

Are total coliforms a good indicator?

If we look back to the WHO requirements, coliforms:

i. grow in drinking water distribution systems;
ii. are normal inhabitants of soil, water and plants; and
iii. are not always present during waterborne disease outbreaks.

This supports the view that they are not a reliable indicator of potential health risks in drinking water.

One of the requirements of a robust indicator of faecal pollution is that it enters the drinking water system with the pollutant and survives for a time that is consistent with the survival of pathogenic micro-organisms. If a water quality indicator can multiply in the environment or in a drinking water distribution system, then detection does not necessarily imply that the system has been compromised by a pollution event or that the water represents a health risk.

Coliform bacteria can grow within biofilms and can be significant contributors to biofilm populations (Chevalier 1990) and the presence of significant numbers of coliform bacteria within distribution systems have been shown not always to index heath risk from associated pathogens.

To support the theory that bacteria growing in drinking water distribution systems do not represent a direct health risk to consumers, a review of the health significance of such bacteria found no evidence of any reported outbreaks of waterborne infection caused by the typical bacteria growing in a distribution system.(PHLS 1994).

Reported waterborne disease outbreaks often demonstrate that total coliforms are not a sensitive indicator of the risk of waterborne disease. Indeed, in some reported outbreaks total coliforms and *E. coli* have been found to be present in others they are absent.

In the USA between 1978 and 1986, there were 502 reported outbreaks of waterborne disease involving more that 110,000 cases of gastrointestinal illness, many of the implicated water supplies passed the coliform standard test. An additional study found that over one third of water supplies responsible for outbreaks of disease did not have total coliforms present. This empirical evidence lead to the conclusion that the presence of coliforms, including *E. coli*, are sometimes a useful indicator for the presence of viruses and bacteria but not for protozoan parasites.

Waterborne disease outbreaks, where total coliforms are detected, are generally attributed to bacteria, viruses or unknown agents (Moore et al, 1994) WHO reported that 88% of such outbreaks had coliforms present in the water supply but, for outbreaks attributed to protozoan parasites only, 33% of waters were positive for coliforms.

Much of the difficulty in correlating the presence of bacterial indicators to the presence of protozoa is due to the differing susceptibility of protozoa to treatment processes, particularly chlorination. Chlorine as a disinfectant is more effective against bacteria and viruses than protozoa, making coliforms limited in their use as a measure of the effectiveness of treatment processes (Sobsey, 1989). Table 1 summarises the issues in application of the coliform-based indicator system in drinking water management.

4 WHAT ARE THE ALTERNATIVES?

The interpretation of coliform presence or absence data which is acquired to address the needs of compliance assessment, does not enable water quality managers or health professionals to make confident decisions about the microbial safety of drinking water.

This end point, compliance driven system needs to be replaced with a complete management system which understands the risks posed, the best way to manage them and specific testing to validate the effectiveness of implementation. This is addressed by the implementation of Drinking Water Safety Plans.

Drinking water supply and system monitoring for bacterial indicators has four major purposes.

i. To identify general faecal contamination of source waters.
ii. To demonstrate that treatment and / or disinfection processes are working efficiently.
iii. To alert for possible in system contamination, through cross connections, ingress, pipe break contamination and contamination during storage.
iv. To monitor for biofilm growth, general system cleanliness and the potential presence of opportunistic pathogens.

Many alternative indicators to total coliforms have been proposed including enterococci, sulphite reducing *Clostridium* spp., *Bacteriodes fragilis*, *Bifidobacteria* spp., bacteriophages and non-microbial indicators such as faecal sterols. Of these, enterococci has gained most acceptance particularly in conjunction with *E. coli* (WHO 1996)

Table 1 *Issues in the application of a coliform-based indicator of risk in water supply systems.*

Information need	Currently use	Issues
Identify faecal contamination of source water	Absence of total coliforms & E.coli	Short term changes missed, aged faecal material may not contain coliforms but other persistent pathogens may be present
Demonstrate treatment efficacy	Absence of total coliforms & E.coli	Sampling frequency generally too low to detect chlorination upsets or treatment failures
Information on system contamination	Presence of total coliforms & E.coli	Cross connections result in high numbers of *E.coli* entering system, coliforms may already be present in biofilms so detecting them can be confusing.
Biofilms, system cleanliness, opportunistic pathogens.	Presence of coliforms	Coliforms may be part of the biofilm community and do not represent the majority of bacteria within biofilms. Some biofilms have few or no coliforms but may contain opportunistic pathogens

5 RISK MANAGEMENT APPROACH

A risk management approach is being applied in the UK via Drinking Water Safety Plans to increase confidence in the safety of drinking water and this could be further developed to end reliance in end point testing. This holistic approach systematically assesses risks throughout a drinking water supply from catchment and source water through to customers tap and identifies how risks can be managed andinterventions put in place to ensure barriers and control measures are working effectively. These Drinking Water Safety Plans assess the integrity of the entire water system and are able to incorporate strategies to deal with day to day management of water quality as well as treatment upsets and failures. Across the world, a number of water companies have adopted the HACCP system (Hazard Analysis and Critical Control Point). HACCP is risk prevention/ risk management system used extensively in the food industry and allows water managers to identify and rank risks within the water supply and establish critical control points where those risks could be managed, this seems the logical next step in the Drinking Water Safety Plan process.

A risk management approach for drinking water includes;

i. end point monitoring to verify that the water supplied to consumers is safe; and
ii. operational monitoring to show that treatment processes are functioning properly and that the distribution system integrity is maintained.

End point monitoring cannot, however, be used as a system control measure, but only as a final verification step in a complete risk management plan. Operational monitoring is a means of assessing system performance and results are used to modify system controls to ensure that processes are working within specification. For this reason, on-line and continuous monitoring for operational purposes is better able to support system management than our current reliance on end point testing.

Parameters used within a risk management approach to monitor and verify water quality should be simple and include a range of measurements that can indicate:

- faecal contamination of source waters;
- treatment efficacy;
- faecal contamination from ingress; and
- water stagnation, biofilm growth, system cleanliness and potential presence of opportunistic pathogens.

It is widely acknowledged that the major threat to public health from drinking water is from microbial contamination with human or animal faeces

Regli et al, 1993 concluded;

"Risk from death from known pathogens in untreated water is 100 to 1000 times greater than the risk of cancer from known disinfection by products in chlorinated drinking water."
And
"Risk of illness from pathogens in untreated surface waters is 10,000 to 1,000,000 times greater than risk of cancer from disinfection by products in chlorinated drinking water."

E. coli is regarded as the most sensitive indicator of faecal pollution, large numbers are present in the gut and the fact they are not generally present in other environments supports their continued use.
To increase confidence when monitoring for faecal contamination, enterococci analysis has been used. These are a group of bacteria that can be regarded as indicative of faecal pollution although some can originate from other habitats.

The enterococci has a number of advantages over total coliforms as indicators, namely that:

- they do not grow in the environment;
- they survive for long periods;
- they are numerous to be detected after significant dilution; and
- rapid and simple methods based on defined substrate technology are available for detection and enumeration and are routinely employed.

Clostridium perfringens also offers potential as an indicator of faecal pollution distant in time and/or space from a monitoring point The spores produced by *Clostridium perfringens* are largely faecal in origin and always present in sewage, their spores are highly resistant in the environment anddo not to reproduce in aquatic environments. There is evidence to show that *Clostridium perfringens* may be a suitable indicator for viruses and pathogenic protozoa when sewage is the suspected cause of contamination. (Payment and Franco 1993). Nonetheless, *Clostridium perfringens* is not generally considered a robust indicator of microbial water quality because they can survive and accumulate in drinking water systems and may be detected long after a pollution event has occurred and far from the source. (WHO 1996). Their preferred application is to aid identification of faecal contamination in sanitary surveys.

However, to date other faecal indicators superior to E. coli and enterococci have not been developed to a point where there are methods that are readily available, inexpensive and suitable for routine use.

The real-time monitoring of water filtration plants requires continuous assessment using criteria such as turbidity, disinfection effectiveness and system cleanliness. Assessment is better achieved by using disinfectant residual, Ct, which is a measure of concentration of disinfectant and contact time, combined with the measurement of heterotrophic bacteria. Water operators can be alerted by sudden changes in these parameters and investigate to determine actions required.

6 CONCLUSIONS

We have outlined why total coliforms, as a faecal indicator, is flawed, and why alternatives are needed. *E. coli* remains a key verification tool but a set of indicators such as those detailed in Table 2, used in conjunction with robust total system risk management is required to increase confidence in the safety of water delivered to consumers.

Hence, moving forward, we could use E. coli as the primary indicator of faecal pollution supported by other measurements such as heterotrophic plate count, Ct, chlorine residual and turbidity to verify treatment and disinfection effectiveness and assess system cleanliness.

Table 2 *Indicator systems for total risk management in water supply systems.*

HAZARD	**INDICATOR**
Faecal contamination of source water	Sanitary survey Turbidity HPC
Treatment efficacy	Total chlorine HPC E.coli
Faecal contamination from ingress	Ammonia Enterococci E.Coli Dissolved oxygen (change) Free chlorine (change) Pressure (sudden change
Water stagnation	Loss of chlorine residual Dissolved oxygen HPC
Potential presence of opportunistic pathogens	HPC Free chlorine

References

WHO (1996) Guidelines for Drinking Water Quality. Second Edition, Volume 2 Health criteria
and other supporting information. World Health Organization, Geneva.

LeChevallier, M.W. (1990) Coliform regrowth in drinking water: a review. *Journal of the American Water Works Association*. 82:74-86.

PHLS (1994) On the Health significance of heterotrophic bacteria growing in water distribution
systems. Public Health Laboratory Services Report number 95/DW/02/1.

Moore, A.C., Herwaldt, B.L., Craun, G.F., Calderon, R.L., Highsmith, A.K. and Juranek, D.D.
(1994) Waterborne disease in the United States, 1991 and 1992. *Journal American Water Works Association*. 86:87-99.

Sobsey, M.D. (1989) Inactivation of health related microorganisms in water by disinfection processes. *Water Science and Technology*. 21(3):179-195.

Regli, S., Berger, P., Macler, B and Haas, C. (1993) Proposed decision tree for management
of risks in drinking water:Consideration for health and socioeconomic factors. In: Safety of Water Disinfection: Balancing Chemical and Microbial Risks. Ed. Gunther F Craun. ILSI Press, Washington DC.

Payment, P. and Franco, E. (1993) *Clostridium perfringens* and somatic coliphages as indicators of the efficiency of drinking water treatment for viruses and protozoan cysts. *Applied and Environmental Microbiology*. 59:2418-2424

IMPROVING BACTERIOLOGICAL WATER QUALITY COMPLIANCE OF DRINKING WATER

Kate Ellis[1,2], Bernadette Ryan[2], Michael R. Templeton[3] and Catherine A. Biggs[1,*]

[1] Department of Chemical and Biological Engineering, The University of Sheffield, Sheffield, UK
[2] Research and Development, Severn Trent Water Ltd., Coventry, UK
[3] Department of Civil and Environmental Engineering, Imperial College London, London, UK
[*] Corresponding author. E-mail: c.biggs@sheffield.ac.uk

1 INTRODUCTION

In the United Kingdom, drinking water quality monitoring for a range of microbiological parameters occurs routinely at water treatment works (WTWs), reservoirs and through randomised point of use (customer tap) sampling. Pathogens are rarely isolated from drinking water due to their low numbers under normal circumstances. For this reason, microbiological quality monitoring focuses on indicator organisms.[1] The indicator organisms that must be measured under European Standards are *Escherichia coli* and enterococci; and in many cases 'additional monitoring requirements' include *Clostridium perfringens* and coliform bacteria.[2] Positive results in these analyses are indicative of environmental or faecal contamination of treated water and all four parameters have prescribed values of zero cells per 100 ml. Heterotrophic bacteria are also enumerated through colony counts at 22 and 37 °C (heterotrophic plate counts, HPCs); this test may be positive without indicating contamination and is used primarily to detect changes in microbiological quality; any 'abnormal change' requires investigation by the water company.[3]

Severn Trent Water operates 141 water treatment works (WTWs), 495 service reservoirs and supplies water to 8 million customers via a network of underground pipes. Routine water quality monitoring for a range of parameters is carried out at WTWs, reservoirs and customers' taps in order to assess compliance with European and national standards. Severn Trent Water has a business target of zero water quality failures and achieves near excellent compliance with the regulations (consistently above 99.9 % since 1997; achieving 99.98 % in 2010).[4] A small number of samples do not meet water quality standards; these are investigated immediately to identify the root cause(s) of the failure and action is taken to ensure quality is restored as soon as possible.

The aims of this project are to understand the reasons for sporadic changes in bacteriological quality and to develop ideas for their prevention. The objectives are to determine the most common causes of failures identified through Severn Trent Water's 'root cause' investigations; to observe the frequency with which investigations were closed with the cause unknown; and to identify the normal fluctuations in water quality at a customer's tap.

2 METHODS

2.1 Identifying the Root Cause of Bacteriological Failures

The details of bacteriological failures from 2009 were requested from Severn Trent Water's Quality and Environment data team for routine analyses from treated waters at WTWs, reservoirs and customers' taps. These failures related to bacteriological testing, following Standard Methods,[3] for coliforms, *E. coli*, *C. perfringens* and enterococci where the result was greater than zero cells per 100 ml. For each failure, the investigative report was downloaded and the failure cause(s) extracted from the report summary page. The results were then classified as single causes, multiple potential causes or unknown cause. The unknown category was sub-divided to show whether it was possible to re-sample from the original failing sample point.

2.2 Tap Water Quality and Internal Surface Cleanliness

2.2.1 Customer Tap Study. Sampling took place at a single property and occurred twice weekly on Monday and Thursday mornings. The only exceptions were on public holidays when the samples were collected in the evening and delivered the following morning to ensure the time between sampling and bacteriological analysis was less than 24 h. The tap was flushed for 2 min prior to sampling. On-site free and total chlorine measurements were made using a Hach Chlorine Pocket Colorimeter™ II (Hach Lange GmbH, Düsseldorf, Germany) and Palintest® DPD Tablets No. 1 and No. 3 for 10 ml samples (Palintest Ltd., Tyne and Wear, UK). Approximately 100 ml of water was collected in an insulated cup and the temperature measured using a nitrogen-filled mercury thermometer with a range 0-40 °C. A 500 ml pre-tap-disinfection water sample was collected in a 500 ml sterile plastic bottle dosed with sodium thiosulphate (NaS_2O_4). The tap was turned off and disinfected with a 10,000 mg Cl l^{-1} solution made up using Instachlor® Rapid Release Chlorine Tablets (Palintest Ltd., Tyne and Wear, UK) and a 2 min contact time. The tap was flushed for a further minute and a 500 ml post-tap-disinfection sample collected in a 500 ml sterile plastic bottle dosed with NaS_2O_4. Samples were transported on ice to Severn Trent Water's microbiology laboratory for analysis: HPCs at 22 and 37 °C, coliforms and *E. coli*, non-coliforms, *C. perfringens* and enterococci. These analyses were carried out using Standard Methods.[3]

2.2.2 Adenosine Tri-Phosphate (ATP) Swab Test. 3M™ Clean-Trace™ Surface ATP (3M™, Minnesota, USA) test swabs were used to swab the internal surfaces of a WTW tap which is normally constantly running but was turned off for this test and a laboratory tap which is used intermittently. The swabs were introduced to the 3M™ Clean-Trace™ Surface ATP vials (3M™, Minnesota, USA) containing surfactants which released ATP from cells in the sample for reaction with luciferin/luciferase. This generated a fluorescent signal which was read in a 3M™ luminometer (3M™, Minnesota, USA) with results given in relative light units (RLU). Since ATP is required by all living cells for maintenance, growth and reproduction,[5] this test provides a rapid assessment of the biological (not necessarily bacteriological) cleanliness of surfaces.

3 RESULTS AND DISCUSSION

3.1 Findings of Root Cause Analyses

In Severn Trent Water's area in 2009, there were 43 bacteriological failures (out of nearly 65,000 bacteriological tests), of which 13 had causes identified (Table 1). Both WTW failures and the reservoir failure with a single root cause were due to contamination at the sample point and nine customer tap failures were attributed to unclean taps. The reservoir failure with multiple potential causes identified contamination at the sample point and ingress into the reservoir as contributing to the failure. Targeted action to prevent future non-compliances is simpler if a single cause is identified. Of the 30 non-compliances with no root cause identified, only 3 resulted from being unable to sample the original sample point. Sample line replacement or reservoir cleaning was instituted where the quality of these particular assets was identified as causative. Equally, guidance on tap cleaning was provided to customers where appropriate. However, no further action could be justified where no cause was identified, and this represents the majority of cases.

These findings are corroborated by both the Drinking Water Inspectorate (DWI) and UK Water Industry Research (UKWIR). The DWI Report for 2009 observed that, across England and Wales, microbiological failures represented a significant proportion of non-compliances.[4] Data compiled by UKWIR demonstrated that bacteriological root cause analyses were frequently inconclusive or closed with the cause unknown.[6]

Table 1 *Frequency of successful root cause allocations from a total of 43 bacteriological failures in 2009.*

		Sample Point	
Root Cause of Failure	*WTW Final*	*Reservoir*	*Customer Tap*
Single	2	1	9
Multiple	0	1	0
Unknown - same point	0	11	16
Unknown - different point	0	0	3

3.2 Tap Water Quality

3.2.1 Customer Tap Study. Results are presented for the first 5 months of the single property customer tap survey: i.e. November 2010 to April 2011; of a 12-month study. During routine sampling, the temperature and free and total chlorine concentrations were measured. The results for these parameters demonstrate a decline in free and total chlorine concentrations with an increase in water temperature (Figure 1). No trend in HPCs was observed with an increase in temperature or a decline in chlorine concentration (Figure 2).

There is a consensus that 0.3 mg l^{-1} free chlorine restricts re-growth in distribution,[7,8,9] and, prior to 11th April 2011, free chlorine was in excess of this concentration. On the 11th April the water temperature was 11.9 °C; representing the peak for this time period (Figure 1). Chlorine is more reactive at higher water temperatures. It is, thus, more effective as a disinfectant but is also consumed quicker in reactions with the bulk water and pipe walls.[10,11] The World Health Organization recommends that water temperature be maintained below 15 °C, with or without disinfectants, to reduce microbiological growth in the distribution system.[12] The temperature did not exceed 15 °C during the first five months (Figure 1). The water temperature and chlorine concentrations during this time period may explain the lack of observable trend with HPCs. Blanch *et al.* and Kerneïs *et al.* concluded that temperature alone was insufficient for predicting colony counts;[13,14] however, Szewzyk *et al.* suggest that this is only true for pathogens.[15] Their conclusions were based on randomised sampling within distribution systems,[13,14] or laboratory-scale survival studies.[15] No long-term temperature and disinfection studies have been completed at a single customer property to assess the importance of these factors when determining appropriate measures for managing bacteriological compliance. Completion of the customer tap study will go some way toward clarifying these relationships.

No coliforms, *E. coli*, *C. perfringens* or enterococci were detected. On two occasions, non-coliforms were enumerated pre- and post- tap-disinfection (3rd March and 28th March 2011; Figure 2); these both occurred 1-2 days after cleaning the tap with ecological washing-up liquid and an abrasive sponge. This suggests that the action of cleaning the tap impacts the quality of water samples. In addition, on two other occasions, HPCs were enumerated on the post-tap-disinfection sample only (13th December 2010 and 13th January 2011; Figure 2). Thus interactions with the tap to improve the quality of water samples may, contrarily, lead to their contamination rather than prevent it. If this is the case, then the current sampling protocol may not result in samples that are truly representative of the water in the distribution system.

3.2.2 ATP Swab Test. The constantly running tap gave a reading of 15 RLU and the laboratory tap a reading of 17,000 RLU. The swab blank had a reading of 15 RLU and readings in excess of 300 RLU are deemed to be contaminated with biological material.[16] Since these taps were at the same site and fed with the same treated water it is likely that the way the taps are used (that is, constantly running or intermittent use) has impacted the biological quality of the surfaces within the taps.

Turbulent flow conditions in pipes have been shown to lead to more rapid biofilm growth and more frequent cell-detachment, or sloughing.[17,18,19] Percival *et al.* observed compressed and strongly adhered biofilms in pressurised pipe systems and Hu *et al.* observed that biofilm thickness was reduced under high velocity conditions.[17,20] It is likely that changes in flow and pressure will also be affective upon biofilms on the internal surfaces of taps experiencing intermittent use.

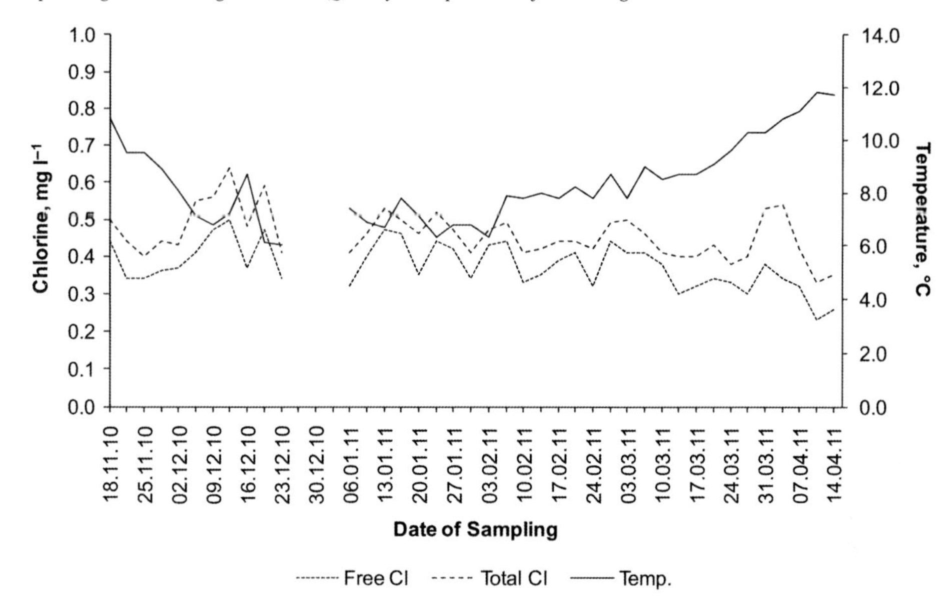

Figure 1 *Temperature and chlorine trends from the Customer Tap Study for the first 5 months of a 12-month investigation; gap = no data.*

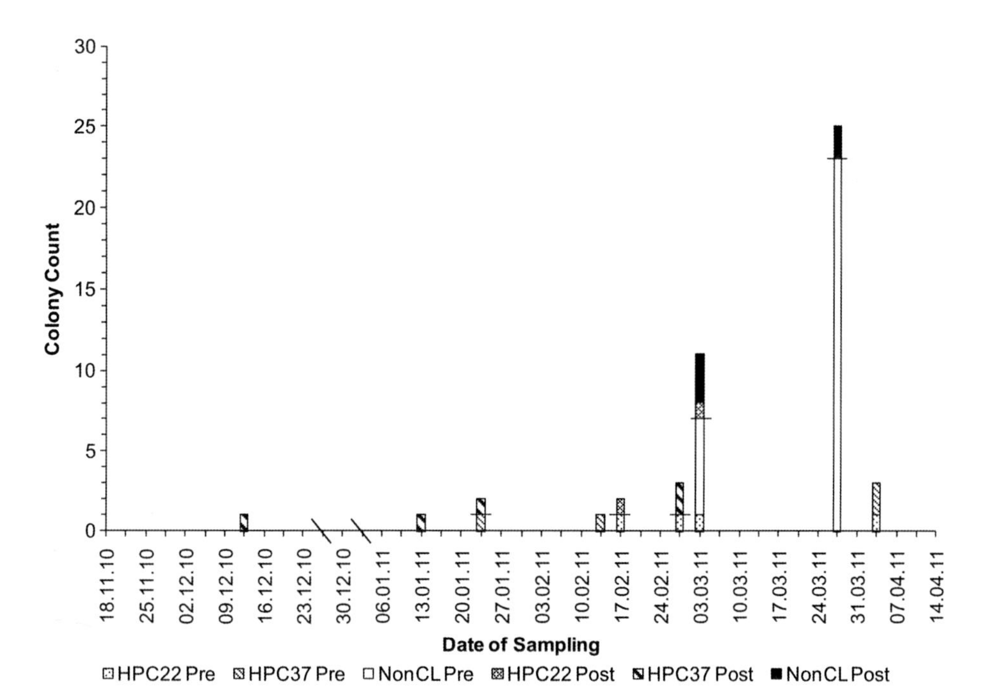

Figure 2 *Bacteriological data from the Customer Tap Study for the first 5 months of a 12 month investigation; \ \ section = no data. Heterotrophic plate counts (HPC) at 22 and 37 °C and non-coliforms (NonCL) pre- and post-tap disinfection.*

4 CONCLUSIONS AND FURTHER WORK

The number of failures for which no cause could be identified represents a significant problem for improving bacteriological compliance. In this study, an ATP test indicated that the manner in which a tap is operated (constantly running or intermittent use) affects the level of biological contamination. Most WTW taps are constantly running, reservoir taps are flushed prior to weekly sampling and customer taps are in intermittent daily use. The customer tap study suggests that disinfecting or cleaning taps de-stabilises the local microbial community and impacts sample quality. Assuming that the initial findings are typical, they go some way to explaining the increased frequency of failures at reservoirs and customer taps and the low incidence of identified causes. Further research is required fully to understand the impact of tap operation and sampling protocols.

An objective of future research is to determine whether sampling from constantly running taps improves the bacteriological quality of water samples compared to sampling from taps that are turned off between samplings. It is hypothesised that quality will be greatest in samples from the constantly running taps, and that quality will improve with flush-time from the flush-first taps. Furthermore, it is believed that internal surface swabs from the flush-first taps will demonstrate lower bacteriological quality due to the aggregation of bacteria when the tap is turned off.

Acknowledgements

This research is funded under the STREAM Industrial Doctorate Centre and is collaborative between EPSRC, Severn Trent Water, The University of Sheffield and Imperial College London. Thanks are expressed to Severn Trent Water staff at Church Wilne Laboratory, Long Eaton and within the Quality and Environment data team, Coventry and Gareth Lang at 3M™ Food Safety, Loughborough. Catherine A. Biggs also wishes to acknowledge the EPSRC for an Advanced Research Fellowship (EP/E053556/01) and further project funding (EP/E036252/1).

References

1 J. N. Lester and J. W. Birkett, *Microbiology and Chemistry for Environmental Scientists and Engineers, 2nd Edn.*, E & FN Spon, London, 1999, ch. 1, pp. 1-11.
2 Council of the European Communities, *Council Directive of 3 November 1998 on the quality of water intended for human consumption. 98/83/EC*, Publications Office of the European Union, 1998.
3 Standing Committee of Analysts, *The Microbiology of Water 1994: Part 1 – Drinking Water*, Her Majesty's Stationery Office, London, 1994.
4 Drinking Water Inspectorate, *http://.dwi.defra.gov.uk/about/annual-report/index.htm*, last up-dated 01 July 2010, accessed 28 February 2011.
5 G. L. Zubay, W. W. Parson and D. E. Vance, *Principles of Biochemistry*, Wm. C. Brown Publishers, Dubuque, 1995, ch. 2, pp. 29-48.
6 UK Water Industry Research, *Validating the Cause of Coliforms in Drinking Water – Final (Draft)*, UK Water Industry Research Limited, London, 2009, pp 6 and 9
7 B. Mahto and S. Goel, *J. Environ. Sci. Eng.*, 2008, **50**, 33
8 A. Francisque, M. J. Rodriguez, L. F. Miranda-Moreno, R. Sadiq and F. Proulx, *Water Res.*, 2009, **43**, 1075

9 Y. Wang, J. Zhang, Z. B. Niu, C. Chien, P. P. Lu and F. Tang, *Water Sci. Technol.: Water Sup.*, 2009, **9**, 349
10 C. N. Sawyer, P. L. McCarty and G. F. Parkin, *Chemistry for Environmental Engineering and Science, Fifth Edn.*, McGraw-Hill, New York, 2003, ch. 20, p. 583
11 S. A. Parsons and B. Jefferson, *Introduction to Potable Water Treatment Processes*, Blackwell Publishing, Oxford, 2006, ch. 9, pp. 128-132
12 World Health Organization, *Safe Piped Water: Managing Microbial Water Quality in Piped Distribution Systems*, ed. R. Ainsworth, IWA Publishing, London, 2004, ch. 1, p. 12
13 A. R. Blanch, B. Galofré, F. Lucena, A. Terradillos, X. Vilanova and F. Ribas, *J. Appl. Microbiol.*, 2007, **102**, 711
14 A. Kerneïs, F. Nakache, A. Deguin and M. Feinberg, *Water Res.*, 1995, **29**, 1719
15 U. Szewzyk, R. Szewzyk, W. Manz and K-H. Schleifer, *Annu. Rev. Microbiol.*, 2000, **54**, 81
16 3M™, *3M™ Clean-Trace™ Surface ATP Protocol*, Personal Communication, Gareth Lang, 3M™ Food Safety, Loughborough, UK, 2011
17 S. L. Percival, J. S. Knapp, R. G. Edyvean and D. S. Wales, *Water Res.*, 1998, **32**, 2187
18 A. O. Al-Jasser, *Water Res.*, 2007, **41**, 387
19 K. Helmi, S. Skraber, C. Gantzer, R. Willame, L. Hoffmann and H-M. Cauchie, *Appl. Environ. Microbiol.*, 2008, **74**, 2079
20 T-L. Hu, C. Lin and W-Y. Chen, *Water Sci. Technol.: Water Sup.*, 2008, **8**, 513

A WATERBORNE OUTBREAK CAUSED BY A SEVERE FAECAL CONTAMINATION OF DISTRIBUTION NETWORK: NOKIA CASE

I.T. Miettinen[1], O. Lepistö[2], T. Pitkänen[1], M. Kuusi[3], L. Maunula[4], J. Laine[3,5], J. Ikonen[1], and M-L. Hänninen[4]

[1]National Institute for Health and Welfare, P.O. Box 95, FI-70101 Kuopio, Finland
[2] Department of Environmental Health Care, Pirkkala community, Suupantie 6, FI-33960, Pirkkala, Finland
[3]National Institute for Health and Welfare, Mannerheimintie 166, FI-00300, Helsinki, Finland
[4] Department of Food Hygiene and Environmental Health, Faculty of Veterinary Medicine, University of Helsinki, P. O. Box 66, FI-00014, Finland
[5]Tampere University Hospital, P.O.Box 2000, FI-33521 Tampere, Finland

1 INTRODUCTION

In 1997, a new notification system for waterborne outbreaks was introduced in Finland. In this system, municipal health protection authorities have to inform national authorities of all suspected waterborne outbreaks. A notification of an outbreak has to be given as soon as possible after a suspicion of an outbreak linked to the quality of drinking water has come out i.e. before confirmative microbiological and chemical analyses of the quality of drinking water have been carried out. The notification which nowadays is an internet based form, is aiming at immediate information on suspected waterborne outbreaks in order to start outbreak investigation and control measures in time. The form is filled by municipal health protection authorities who, in Finland, are responsible for frequent surveillance of the quality of drinking water. The National Institute for Health and Welfare (THL) manages waterborne outbreaks by maintaining a national task group, which helps local authorities in technical, analytical and epidemiological problems associated with outbreaks. Between 1998 and 2010, 69 outbreaks resulting in 27,300 illness cases were notified (unpublished results). Outbreaks have typically been associated with small ground water supplies or private wells serving less than 500 consumers. Typically, these water abstraction plants utilize minimal water treatment. Norovirus and *Campylobacter jejuni* have been the most common microbes causing waterborne outbreaks in Finland[1].

2 THE OUTBREAK

The technical survey of the outbreak revealed a massive contamination of the drinking water distribution network at Nokia. This was due to the cross-connection between cleaned waste water and tap water pipelines at the waste water treatment plant in November-December 2007. The connection was open for two and half days resulting in intrusion of 450 m^3 of cleaned waste water into the drinking water distribution line. The cleaned waste water (waste water was cleaned with chemical precipitation and activated sludge treatment) had been used for at least two decades in the waste water plant for cleaning purposes and as a chemical dilution solution in the waste water treatment process. The waste water plant is located in the middle of Nokia city (total population 30,000 inhabitants).

Among the early signals of contamination were customer complaints beginning on 28th November reporting poor drinking water quality including abnormal color, taste, smell, and foam formation in tap water. The first complaints were directed to the Nokia waterworks. The control room of the water works received several hundred complaints during the first few days of the contamination. No immediate actions were made because the administration of the waterworks assumed the abnormal water quality was caused by deposits loosend from the old, network by changes in water flow direction in the distribution network due to unusually intensive operational factprs.

The first patients with gastroenteritis sought medical treatment on 30th November in the local health care centre. The symptoms included nausea, vomiting, diarrhoea and fever. This led to water sampling for indicator bacteria. Once information about water quality complaints, which seemed to be related to the gastroenteritis, reached the local health inspector, the outbreak was finally recognized and mitigation actions to stop the outbreak were initiated. Extremely high counts of indicator bacteria, *E. coli* and intestinal enterococci, were identified from the first tap water samples taken from the network. The first water samples also had high turbidity (turbidity decreased from 4.6 to 0.2 NTU within 16 days) and particle counts . Later, *Campylobacter sp.*, noroviruses, *Salmonella* sp., and rotaviruses were detected both from water and human faecal samples[3,4]. In addition, entero-, astro- , rota- and adenoviruses were detected from water samples[3]. Also *Giardia* spp. were identified when *Giardia* cysts were detected from one deposit sample taken from the drinking water distribution network at the beginning of December 2007[5].

3 MITIGATION MEASURES

The staff of the waterworks were able to locate the source of contamination at the waste water plant on the same day as the contamination was finally recognized. The faulty connection between the wastewater line and tap water line was immediately rectified and flushing of the network with clean water was commenced. Based on analytical measurements and hydraulic information, it was possible to define the contaminated area and decontamination actions were deployed over an area with 9,538 inhabitants. Normally, Nokia city had disinfected the distributed drinking water with chlorination (approx. 0.3 mg/l). One of the first mitigation actions was to elevate chlorine concentration to 1.5 mg/l and later it was again raised to 3 mg/l. Although, the indicator organisms disappeared during the first week, pathogenic microbes (viruses) were found for over two months in samples from the network. It seemed that even enhanced chlorination was insufficient to clean the badly contaminated pipelines. This resulted in the use of internal cleaning (pigging) and air/water pulsed flushing to clean the main pipelines. Additionally, shock chlorination (10 mg/l, 24 h) was used to enhance cleaning of especially the connection lines and the indoor pipeline installations. Eighteen alternative drinking water delivering service points (water tanks and bottled water) were also opened in the contaminated city area during the outbreak.

A boiling water advisory (5 min boiling time) of drinking water was ordered on the day the outbreak was confirmed. Information about the drinking water contamination and boiling water advisory was communicated using the local radio channels, Internet, TV and newspapers. Leaflets and loudspeaker cars were used to inform customers. Also, schools, food factories and hospitals were informed to prevent drinking of contaminated (un-boiled) water. Due to the persistent incidence of pathogens in the contaminated network, the boil water advisory continued for almost three months. It was cancelled gradually after all cleaning measures in February 2008 when neither noroviruses or adenoviruses could be found in water samples taken from the network.

4 OUTCOME OF THE OUTBREAK

A population survey was conducted in Nokia and neighbouring towns to explore the true extent of the outbreak and to evaluate its consequences. The outcome of the study was that a total of 8,453 cases of gastroenteritis occurred during the outbreak, the excess number of illnesses being 6,500. 1222 patients visited the municipal health centre and nearly 200 needed medical care in the hospital[2]. Service points, offered an alternative drinking water supply distribution facility. Some 5,000,000 litres of tank water and 700,000 litres of bottled water were supplied to customers during the outbreak. The total costs of the outbreak, including the technical mitigation actions, reimbursed hospital costs, personal claims for damages, water analyses, and extra workload exceeded 4.6 million euros.

This was the largest waterborne outbreak in Finland for at least 20 years and it affected strongly the national water services. It was unusual to face such a large-scale, drastic waterborne outbreak which included several intestinal pathogens. This was also the first waterborne outbreak associated with drinking water contaminated by *Giardia* spp. for over 15 years in Finland[5]. One of the crucial lessons of this outbreak's management was the lack of information describing the first signs of the contamination. This resulted in wide distribution of faecal pathogens along the distribution system. In fact, the *E. coli* counts in the drinking water were so high that they would have been detected early (after a couple of hours) by utilizing a broth culture based on fluorescence detection such as Colilert®. Also, physical on-line meters of water quality could have warned of the contamination if they had been deployed by the water undertaker. On the other hand, this outbreak showed how the culturability of the indicator bacteria decreases quite quickly after introduction in to the drinking water system, thus, lowering their value as indicators of initial risk and decontamination efficiency. In the Nokia example, the removal of noroviruses and adenoviruses were used to confirm the cleaning efficiency of networks.

As experienced in many earlier waterborne outbreaks, successful public communication was extremely challenging to deliver. Earlier recognition and informing of the contamination could have limited the extent of the disease outbreak. More efficient information communication, especially at the beginning of the outbreak, would have reduced the customers′ criticisms concerning the capability of authorities to manage the outbreak. As a result of the outbreak, a national survey was launched by the Finnish Ministry of Social Affairs and Health to find and remove any other waste water and clean water pipeline cross-connections in Finland to prevent similar outbreaks to that which occurred at Nokia.

References

1. Miettinen IT, Zacheus O. Pitkänen T., Kuusi M., Vartiainen T. 2005. Proceedings of *15th International Symposium on Health Related Water Microbiology*, Swansea, UK. 5-9.9.2005, pp. 39-40.

2. Laine J, Huovinen E, Virtanen MJ, Snellman M, Lumio J, Ruutu P, Kujansuu E, Vuento R, Pitkänen T, Miettinen IT, Herrala J, Lepistö O, Antonen J, Helenius J, Hänninen ML, Maunula L, Mustonen J, Kuusi M, and the Pirkanmaa Waterborne Outbreak Study Group. 2010. *Epidemiol. Infect.* 2011; **139**: 1105-1113.

3. Maunula L, P. Klemola, A. Kauppinen, K. Söderberg, T. Nguyen, T. Pitkänen, S. Kaijalainen, M. L. Simonen, I. T. Miettinen, M. Lappalainen, J. Laine, R. Vuento, M. Kuusi and M. Roivainen. 2009. *Food Environ. Virol.***1** (1): 31-36.

4. Pitkänen T, Maunula L, Hänninen M-L, Siitonen A, Lepistö O, Mäkiranta M, Kuusi M, Miettinen IT. 2011x. In preparation.

5. Rimhanen-Finne R, Hänninen M-L, Vuento R, Laine J, Jokiranta ST. Snellman M, Pitkänen T, Miettinen IT, and Kuusi M. 2010. *Scan. J. Infect. Dis.*, **42** (8): 613-619.

OCCURRENCE AND GROWTH OF COLIFORM BACTERIA IN DRINKING WATER DISTRIBUTION SYSTEMS

B. Hambsch[1], A. Korth[2], H. Petzoldt[2]

[1] DVGW Technologiezentrum Wasser (TZW), Karlsruher Strasse 84, 76139 Karlsruhe
[2] DVGW Technologiezentrum Wasser (TZW), Aussenstelle Dresden, Wasserwerkstrasse 2, 01326 Dresden

1 INTRODUCTION

In Germany, most distribution systems are run without disinfectant residuals. At the same time, the drinking water directive gives a limit value for *E. coli* and for total coliform bacteria (not detectable in 100 mL). To allow an adequate judgement of coliform-positive samples in the distribution system, it was important to know if growth of coliform bacteria is to be expected in drinking water. Therefore, the growth conditions for coliform bacteria in drinking water were characterized in a DVGW funded project.

In some countries (e.g. the US or Australia) an increase in coliform bacterial numbers during water distribution had been observed and was correlated to certain parameters of the system, like a high level of assimilable organic carbon (AOC > 100 µg/L acetate-equivalents), a temperature higher than 15°C and a low level of chlorine residual.[1,2,3,4] In Germany, the AOC-concentration is usually much lower then 100 µg/L and, in most cases, the water temperature is below 10°C. Thus, coliform bacteria were not expected to grow under the conditions present in German drinking water (i.e. low temperatures, low nutrient contents). In a study by Szewzyk and Conradi (2003)[5] it was also concluded that no growth of coliform bacteria occurs in water or biofilms within drinking water systems.

Nevertheless, coliform bacteria are sometimes detected, for instance on plastic materials[6] or in sediments[7]. Also under practice conditions, the authors found coliform bacteria e.g. in flushing waters as well as in sediments.

Therefore, it was important to characterize the growth conditions for coliform bacteria in more detail for the situation in Germany. This was done within a DVGW-funded project. In this project, a representative study on the presence of coliform bacteria in flushing waters and sediments of reservoirs was completed. Furthermore, the growth conditions for coliform bacteria were analyzed for water, biofilm and sediments under laboratory conditions.

2 PRESENCE OF COLIFORM BACTERIA IN SEDIMENTS AND FLUSHING WATER

Faecal contaminations in German drinking waters are published in official reports.[8] In the years 2002-2004, the percentage of *E. coli*-positive samples was between 0.11 and 0.23 %. For coliform bacteria, the percentage was slightly higher (about 1 %). In the Micro Risk project (EVK1-CT-2002-00123), the background concentration of *E. coli* was analysed by sampling large volumes of water from water supplies in Germany, the Netherlands and the UK. As no *E. coli* cells could be detected, it could be concluded that such a background contamination level is either not present or below 8×10^{-5} CFU/100 mL.[9] Consequently, the occasional presence of *E. coli* should be the result of a contamination event. But the presence of coliform bacteria was also reported for sediments.[7] Therefore, within this DVGW-funded study, flushing waters and sediments in reservoirs were analysed for coliform bacteria in about 10 different distribution systems.

Overall, 122 samples of flushing waters were analysed for the presence of coliform bacteria. The results are summarized in Table 1. In 6 of the 10 distribution systems a part of the samples was positive for coliform bacteria. The percentage of positive samples per system ranged between 15 and 65% with absolute concentration up to 90 coliform bacteria per 100 mL. Mainly *Enterobacter*-species (*E. cloacae, E. agglomerans, E. sakazakii, E. amnigenus*), *Citrobacter*-species (*C. freundii, C. youngae*), *Klebsiella ozeanae* and *Butiauxella agrestis* could be identified (API 20E system) from these samples..

Table 1 *Samples during flushing procedures in 10 distribution systems*

distribution system	coliform bacteria			
	positive	total	%	C_{max}/100 mL
A	2	14	14	2-4
B	4	14	29	1-90
C	4	6	67	1-20
D	3	9	33	1-4
E	1	5	20	2
F	2	6	33	1-10
G	0	6	0	0
H	0	3	0	0
I	0	4	0	0
J	0	3	0	0

Sediments from 9 reservoirs were also analysed for coliform bacteria (Table 2). Coliform bacteria could be detected in the sediments of 7 out of 9 reservoirs, with concentrations varying in a wide range from 35 up to 45000 coliform bacteria per g dry matter. *E. coli* were not detected at all.

Table 2 *E. coli and total coliform bacteria in sediments of reservoirs*

reservoir	raw water*	coliform bacteria/*E. coli* per g dry matter
1	sw + bf	45000/0
2	sw	1920/0
3	sw	89/0
4	sw	43/0
5	sw	35/0
6	gw + bf	6/0
7	gw	1/0
8	sw	0/0
9	sw	0/0

*sw: surface water; bf: bankfiltrate; gw: ground water

3 GROWTH EXPERIMENTS WITH COLIFORM BACTERIA IN LABORATORY SCALE

The batch experiments in the water phase were performed with drinking water produced form surface water (i.e. following treatment involving flocculation, filtration and disinfection), with model water and with water samples collected during pipe flushing. The water samples were inoculated with the coliform strains *Enterobacter asburiae, Citrobacter freundii* and *Enterobacter cloacae* and with a natural autochthonous microbiocoenosis.

Pure cultures of *E. cloacae* (DSM 30054), *C. freundii* (DSM 30039) and *E. asburiae* (isolated from a drinking water reservoir) were grown on nutrient agar. The colonies were transferred to 0.9% NaCl-solution and homogenized. After filtration through a 0.2 μm membrane filter the bacteria were resuspended in 0.9% NaCl-solution. The natural biocoenosis was obtained by filtration of 40 L drinking water through a 0.2 μm membrane filter and resuspending the bacteria in 0.9% NaCl-solution. Additionally, inorganic nutrients and sodium acetate, as an organic substrate, were added.

The samples were incubated at 22 °C and 12 °C on a shaker at 100 rpm to ensure sufficient oxygen supply. Over a 3-4 weeks period, samples were taken regularly to determine the growth of the coliform bacteria and the microbiocoenosis. Coliform bacteria were enumerated by cultivation on Chromocult® Coliform Agar. The growth of the microbiocoenosis was determined by using R2A-Agar.

The growth experiments showed that pure cultures of all 3 different coliform strains could multiply in model water as well as in real drinking water at the incubation temperature of 22 °C. Their growth was depending on the added acetate concentration, but also without addition of organic substrate, a considerable increase in cell numbers (up to 10^7/L) was observed. At a temperature of 12 °C, no growth of *E. cloacae* in pure culture was observed, while *E. asburiae* and *C. freundii* could still multiply at this lower temperature. So the temperature conditions for multiplication depend strongly on the species.

In another experimental setup, coliform bacteria were added together with a natural microbiocoenosis to the water samples. The growth of the natural microbiocoenosis was much faster than the growth of the added coliform strains and, therefore, no multiplication of the coliform bacteria in these experiments occurred, regardless which strain was used.

An example is given in Figure 1, showing the growth of *E. asburiae* and the mixed microbiocoenosis in drinking water produced from surface water at 22 °C. *E. asburiae* could multiply in pure culture, reaching the growth maximum after 3 weeks. In the presence of the mixed microbiocoenosis, *E. asburiae* did not multiply at all (Figure 1, left hand side) whereas the mixed microbiocoenosis multiplied in less than 1 week to the maximum numbers (Figure 1, right hand side). Thus, the natural microbiocoenosis out-competes the coliform bacteria. None of the 3 coliform species grew in the presence of the natural microbiocoenosis in drinking waters and model waters. Such a competition for substrates was also suggested from growth kinetics by Van der Kooij, 1997[10].

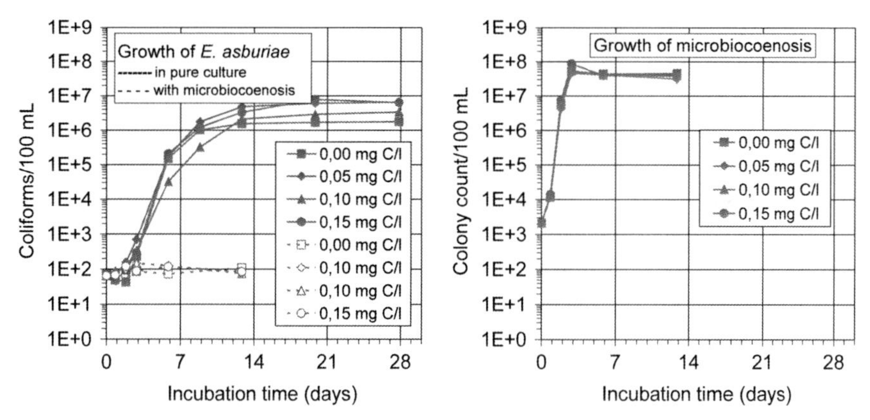

Figure 1 *Growth of coliform bacteria and autochthonous microbiocoenosis in drinking water produced from surface water (with added sodium acetate and inorganic nutrients) after inoculation with Enterobacter asburiae and the natural microbiocoenosis*

As coliform bacteria could be detected also in flushing waters and sediments, another set of experiments was completed with organic matter from these waters or sediments as substrates. The results of experiments in water samples from pipe flushing are shown in Figure 2. *Enterobacter cloacae* was added to these water samples including the mixed microbiocoenosis either directly or after autoclaving (= without microbiocoenosis). The coliform concentration did not increase in the untreated flushing water sample, but did increase in the autoclaved sample (Figure 2, left hand side). The increase of the mixed microbiocoenosis is shown on the right hand side of Figure 2. The experiment showed that also with organic substrates from flushing waters the presence of a natural microbiocenosis prevents growth of coliform bacteria. Yet again, pure cultures of coliform bacteria are shown to multiply with such organic substances.

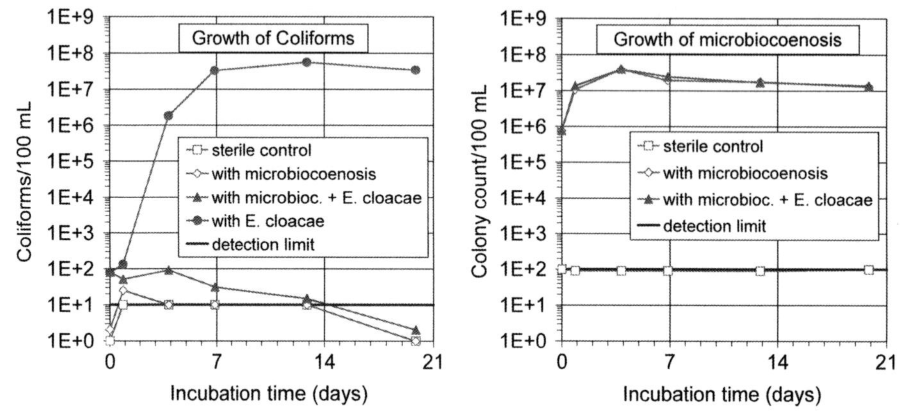

Figure 2 *Growth of coliform bacteria and autochthonous microbiocoenosis in water samples (collected during pipe flushing) and inoculated with Enterobacter cloacae and microbiocoenosis after autoclaving and addition of inorganic nutrients.*

Another set of batch experiments was completed with sediments from drinking water reservoirs, a very high sediment concentration (500 g/L) and a lower concentration (10 g/L) were employed. Both batches were inoculated with *Enterobacter asburiae*. The results of the enumeration of the coliform bacteria and the mixed microbiocoenosis (HPC) as well as the oxygen concentration are shown in Figure 3, on top the higher sediment concentration, on the bottom the lower sediment concentration. In the case of high sediment concentration, oxygen was depleted by chemical and bacteriological processes to reach final levels below 1 mg/L after 2 weeks. In contrast, in the experiment with low sediment concentration, no considerable O_2-consumption was seen (> 6 mg/L O_2 after 2 weeks).

Growth of coliform bacteria was only detected in the batch with the high sediment concentration, whereas with the low concentration the number of coliform bacteria did not increase. From these results, and some other more detailed experiments, it could be concluded that, at very high substrate concentrations and under anaerobic conditions, the multiplication of coliform bacteria occurs even in the presence of a mixed natural microbiocoenosis. Thus, such conditions are of advantage for the facultatively anaerobic coliform bacteria compared to the other mainly aerobic heterotrophic bacteria. These results also explain the presence of coliform bacteria in sediments (see 2), which is also reported in the literature, further highlighting the importance of sediments as growth environments for coliform bacteria.[1, 11, 12]

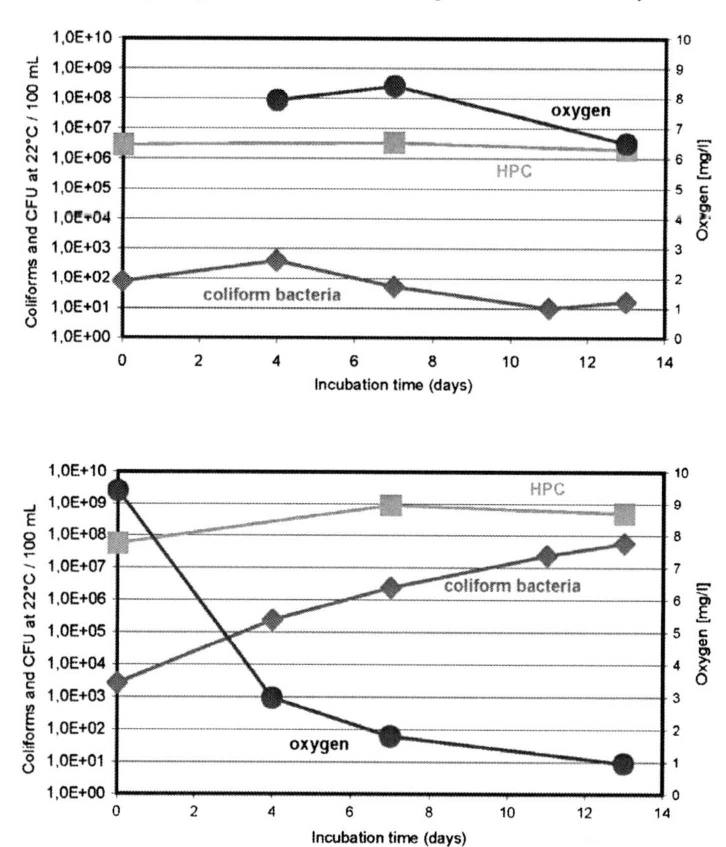

Figure 3 *Coliform bacteria, heterotrophic plate counts (HPC) and oxygen in batch experiments with sediments (500 g/L top, 10 g/L bottom).*

4 COLIFORM BACTERIA IN BIOFILMS

The growth of coliform bacteria in biofilms was investigated in laboratory experiments by adding different *Enterobacter* species to coupons covered with a natural drinking water biofilm. The strains *E. cloacae*, *E. asburiae* and *E. amnigenus* were added and the coupons were incubated at different temperatures and substrate concentrations. After the introduction of the coliform species into the biofilm, only a very small part of the biofilm population was identified as coliform bacteria (mostly below 1 %). Moreover, after the introduction step, the coliform bacteria did not increase in cell number, and within a few days the coliform bacteria were not detected in the biofilm.

A bench scale experiment in a water works, under flow-through conditions, yielded the same results (Figure 4). Coliform bacteria were detected in the biofilm as long as they were present in the water phase. After stopping the dosage of coliform bacteria into the water, they did not persist in the biofilm. Thus, it can be concluded that a natural biofilm in drinking water systems does not offer favourable conditions for the growth of coliform bacteria. This might not be the outcome if materials are used that allow the formation of very thick biofilms. Under such conditions, anaerobic patches could occur within the biofilm, giving a growth advantage for coliform bacteria.

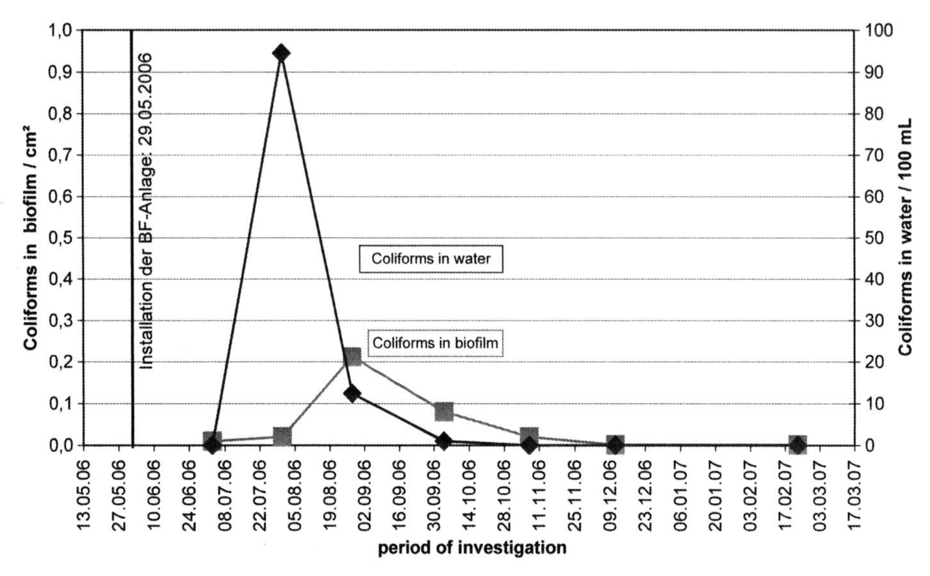

Figure 4 *Coliform bacteria in a flow-through column in a water works in water and biofilm*

5 DISCUSSION AND CONCLUSIONS

The study presented here showed that pure cultures of coliform bacteria can multiply also in the low substrate concentrations of drinking water in Germany in the µg C/L-range. All tested strains did grow at 22 °C; some strains were also able to multiply at 10 – 12 °C. Growth tests in the presence of a natural microbiocoenosis revealed that these indigenous heterotrophic bacteria successfully out-compete coliform bacteria, confirming kinetic observations from Van der Kooij, 1997[10].

Moreover, experiments with sediments showed that, for high substrate concentrations and anaerobic conditions, coliform bacteria are competitive compared to the indigenous bacteria and, therefore, can multiply and survive for a long period of time. Whereas, under aerobic conditions, the substrate uptake by the natural microbiocoenosis usually is much faster and consequently coliform bacteria can not multiply.

Furthermore, in natural drinking water biofilms, no growth of coliform bacteria was observed. This may not be the case if materials or products are present that produce very intense biofilms with anaerobic microenvironments. The results of laboratory experiments confirm and explain the findings of field studies on sediments and problems in water distribution systems, where coliform bacteria are often detected in sediments and on materials with extreme biofilms. As soon as high substrate concentrations and anaerobic conditions occur, survival and multiplication of coliform bacteria is probable, even at low temperatures around 10 °C.

The results of this study agree with most of the results in the international literature, but some have to be interpreted in light of the specific conditions in German distribution systems.

Coliform bacteria are not a conclusive indication of faecal intrusion into the system, as they can survive and multiply, for instance, in sediments. But a multiplication of coliform bacteria is not to be expected under the normal conditions prevalent in German drinking water distribution systems which is characterized by: (i) the presence of natural microbiocoenosis which outcompetes the coliform bacteria; (ii) aerobic conditions in water and biofilm; and (iii) low nutrient conditions in the water.

As growth of coliform bacteria can occur under anaerobic conditions, sediments or extreme biofilms on materials are potential habitats for coliform bacteria. Some of the coliform positive samples in distribution systems with generally low numbers of coliform bacteria are probably caused by the mobilization of sediments due to sudden increases in flow.

The suggestions for drinking water utilities seeking to avoid coliform positive samples during distribution are therefore:

a. to prevent the intrusion of coliform bacteria by distributing a drinking water without coliform bacteria and preventing ingress during distribution;
b. avoid the build-up of sediments in the distribution systems by regular and effective flushing programs;
c. maintain a more or less constant flow regime under normal conditions; and
d. avoid use of materials that bring about extreme biofilm formation, in which also coliform bacteria might find a potential habitat.

Most of these suggested actions are already enshrined in acknowledged rules applied in Germany. Regarding the presence of coliform bacteria in German drinking water distribution systems, it was decided to put this parameter to the 'indicator parameters' of the German Drinking Water Directive which implies less stringent remedial action by authorities than would be the case if it were considered a 'microbiological' parameter.

Acknowledgement

This study was funded by the German Association of Gas and Water (DVGW) within the project W6/03/04.

References

1 A.K. Camper, G.A. McFeters, W.G. Characklis, W. L.Jones, Growth kinetics of coliform bacteria under conditions relevant to drinking water distribution systems, *Appl. Environ. Microbiol.,* 1991, **57**, 2233.
2 K.N. Power, L.N. Nagy: Relationship between regrowth and some physical and chemical parameters within Sydney's drinking water distribution system. Water Research, 1999, **33**, 741.
3 C.J. Volk, M.W. Le Chevallier, Assessing biodegradable organic matter, *Journal AWWA,* 2000, **95**, 64.
4 M. Boulam, S. Fass, S. Saby, V. Lahoussine, J. Cavard, D. Gatel, L. Mathieu, Organic matter quality and survival of coliforms in low-nutritive waters, *Journal AWWA*, 2000, **95**, 119.
5 U. Szewzyk, B. Conradi, Mikrobielle Ökologie in Trinkwasser-Biofilmen,*In:* Berichte aus dem IWW, 2003, **36**, 232.
6 B. Kilb, B. Lange, Trinkwasserkontamination durch Biofilme auf weich dichtenden Absperrschiebern, *bbr* 2001, **7**, 55.

7 H. Schreiber, Invertebraten in Wasserversorgungsanlagen: eine Bestandsaufnahme unter hygienischen Aspekten, 1996, *Dissertation an der Rheinischen Friedrich-Wilhelms-Universität Bonn.*
8 Anonymous, *Bericht des Bundesministeriums für Gesundheit und des Umweltbundesamtes an die Verbraucherinnen und Verbraucher über die Qualität von Wasser für den menschlichen Gebrauch (Trinkwasser) in Deutschland,* 2005, Eigenverlag Bonn / Dessau
9 B. Hambsch, K. Böckle, J. H.M. van Lieverloo, Incidence of faecal contaminations in chlorinated and non-chlorinated distribution systems of neighbouring European countries, *Journal of Water and Health*, 2007, **05.Suppl 1**, 119
10 D. van der Kooij, Multiplication of coliforms at very low concentrations of substrates in tap water. In: D. Kay, C. Fricker (ed.) *Coliforms and E. coli Problem or Solution?*, 1997, 195.
11 R.S. Martin, W.H. Gates, R.S. Tobin, D. Grantham, R. Sumarah, P. Wolfe, P.J. Forestall, *Journal American Water Works Association,* 1982, **74,** 34.
12 D.S. Herson, B. McGonigle, M.A. Payer, K.H. Baker, Attachment as a factor in the protection of Enterobacter cloacae from chlorination, *Appl. Environ.Microbiol.,* 1987, **53,** 1178.

PREDICTIVE MODEL OF CHLORINE DYNAMICS IN WATER

D. Kim, C.T. Le, V.V. Ha, D. Frauchiger, A. Doyen, N. Garg

Innovation Centre, Vestergaard Frandsen, Chemin de Messidor 5-7, CH-1006, Lausanne, Switzerland

1 INTRODUCTION

Millions of deaths occur annually in developing countries due to diarrheal diseases caused by poor water quality. The annual burden of diarrheal disease is estimated at 3.5 billion episodes and results in 1.8 million deaths in children worldwide[1,2]. Diarrheal diseases caused by poor water quality are also responsible for school and work absenteeism, and can result in substantial economic losses for families[3]. Protection of water sources and appropriate disposal of human and animal waste is, therefore, necessary to decrease the morbidity/mortality patterns associated with contaminated drinking water.

In under-resourced communities, household water treatment interventions have been shown to improve water quality and reduce diarrheal disease incidence[4]. Devices that are available for reducing microbial content in drinking water include chlorination, solar disinfection, filtration, or combined flocculation and disinfection[4]. Chlorination is an intervention that has been widely used to treat contaminated drinking water. The benefits of chlorination include scalability and low cost; however, major drawbacks include the requirement to maintain constant quality control, its limited effectiveness against parasites, the strong odours and negative taste, and the presence of disinfection by-products that have carcinogenic potential[5,6].

The objective of the current study is to develop a mathematical model of chlorine concentrations in water of varying quality. Mathematical models have been used widely to predict the reaction and transport of chemicals in water distribution systems[7]. Such models make use of systems of differential equations to predict the concentration time-course profiles of chemicals in water. In the study reported here, a mathematical model is constructed and the parameters optimized against laboratory data on chlorine measurements in water with varying turbidity and humic acid levels.

2 METHODS AND RESULTS

2.1 Description of the Mathematical Model

The mathematical model (Figure 1) consists of two compartments described by systems of linear differential equations.

$$\frac{dC_{Cl}}{dt} = k_r C_{St} - (k_s + k_e) C_{Cl} \quad [1]$$

$$\frac{dC_{St}}{dt} = k_s C_{Cl} - k_r C_{St} \quad [2]$$

The assumptions of the model are: (1) the volume of the container is fixed; (2) there is no spatial dependence of chlorine in the water system. The central compartment, *Cl*, contains the measured chlorine concentration (C_{Cl}). The storage compartment, *St*, contains chlorine in storage (C_{St}). The rate constants are k_e for elimination of chlorine from the central compartment (L/h), k_s is the transfer rate from central to storage compartment (L/h), and k_r is the transfer rate from storage to central compartment (L/h). The model code is written in Berkeley Madonna Software (Berkeley, CA).

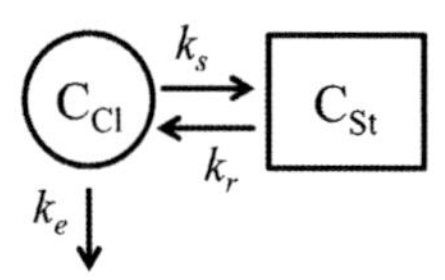

Figure 1 *Mathematical model of chlorine depletion. Chlorine stored at rate k_s and released from the storage compartment at rate k_r; depletion of chlorine occurs in the central compartment at the rate k_e. Two ordinary differential equations are used to describe the dynamic behaviour of the system.*

2.2 Experimental Design

Laboratory experiments were conducted to measure the decrease of residual chlorine concentration as a function of time in aging water. The tests were conducted using water with variable chlorine concentrations, turbidity levels, and humic acid concentrations. The temperature of the water was kept between 25 °C and 26 °C. A spectrophotometer (Hach DR5000) was used to quantify the residual chlorine over time. Measurements were made at time 0, 0.75, 1.75, 4.0, 5.25, 6.25, 7.25, 8.25, 23.25, and 24.75 h.

Table 1 *Experimental conditions for laboratory-based chlorine depletion study.*

Experimental parameter	Values
Initial chlorine concentration	2, 4, and 6 mg/L
Turbidity	5, 20, and 100 NTU
Humic acid	5 and 10 mg/L

2.3 Parameter Optimization

The parameters (i.e., k_e, k_s, k_r) in the mathematical model were optimized against laboratory concentration time-course profiles of chlorine using Berkeley Madonna software. The default curve-fitting method was used, with tolerance set to 0.01, to minimize the deviation between the model's output and the laboratory data. A deterministic fit was made to each set of experimental data (i.e., chlorine concentration and turbidity level). Model fit was evaluated by visual inspection and by minimizing the error between model prediction and observed value.

The model predictions against experimental data are shown in Figure 2. Overall, the two-compartment mathematical model fits the data well. The significance of the storage compartment is revealed in the 2 mg/L initial chlorine concentration data. The concentration time-course profiles show a bi-phasic depletion of chlorine, suggesting that, at lower concentrations, chlorine is bound in a storage compartment. This is confirmed by the observation that the ratio k_r:k_s is equal to or less than 1 across all turbidity levels. Conversely, the ratio k_r:k_s is greater than one for the 4 mg/L and 6 mg/L solutions. All optimized parameter values are reported in Table 2.

The impact of humic acid on chlorine depletion is shown in Figure 3. The presence of humic acid results in more rapid depletion of chlorine from the central compartment; optimal parameters values are approximately two-fold greater in the solution containing 10 mg/L of humic acid versus 5 mg/L, suggesting more rapid transfer of chlorine between the central and storage compartments.

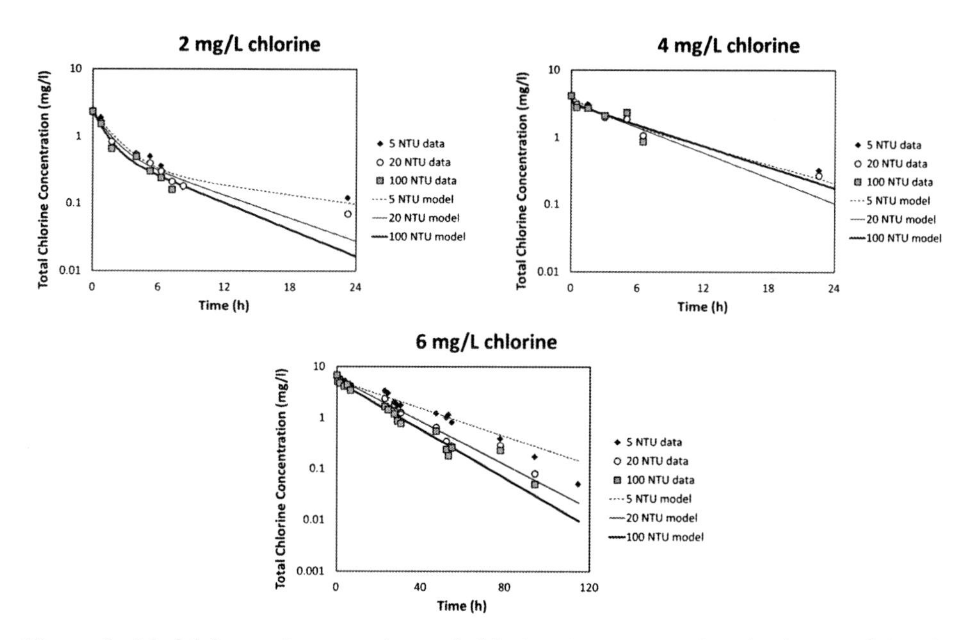

Figure 2 *Model fits against experimental chlorine time-course data for 2, 4, and 6 mg/L initial concentrations. At each chlorine concentration, the turbidity was changed beginning with 5, 20, and 100 NTUs. Humic acid concentration was kept constant at 10 mg/L.*

Table 2 *Optimal parameter values of the chlorine mathematical model against experimental concentration time-course data. Humic acid concentration was kept constant at 10 mg/L.*

Chlorine level (mg/L)	Turbidity (NTU)	k_e (L/h)	k_s (L/h)	k_r (L/h)
2	5	0.23	0.24	0.12
	20	0.32	0.24	0.29
	100	0.39	0.32	0.35
4	5	0.15	0.14	0.43
	20	0.18	3.1	13
	100	0.16	3.8	12
6	5	0.037	0.094	0.54
	20	0.055	1.8	14
	100	0.071	5.5	19

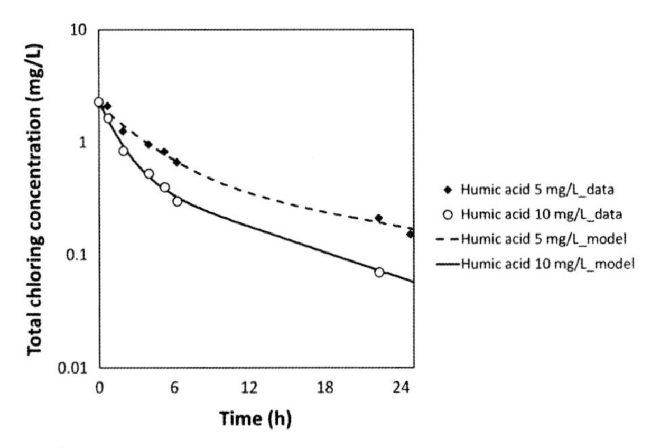

Figure 3 *Model fit against chlorine time-course data under varying humic acid concentrations: 5 mg/L and 10 mg/L. The initial concentration of chlorine was 2 mg/L and the turbidity level was set constant at 20 NTU.*

3 CONCLUSION

Mathematical models are valuable for quantitatively understanding the behaviour of dynamical systems. Chlorine in water is one such dynamical system for which the concentration of chlorine is impacted by the quality (i.e., turbidity level and humic acid content) of water. Our modelling results reveal that the relative impact of turbidity and humic acid is greater in water bodies containing lower levels of chlorine. These findings suggest that rigorous maintenance of an optimal level of chlorine is needed in chemically treated water bodies. This may be achieved through diligent water quality monitoring programs. Though the current work has focused on household water treatment, the mathematical model may also be applied for pool and grey water treatment programs.

References

1. World Health Organization, 2004. *Water, Sanitation, and Hygiene Links to Health.* Available at: www.who.int/entity/water_sanitation_health/publications/facts2004/en.
2. M Kosek, C Bern, R Guerrant, 2003. The global burden of diarrhoeal disease, as estimated from studies published between 1992 and 2000. *Bull World Health Organ* **81**: 197–198.
3. T Clasen, 2008. *Water Quality Interventions to Prevent Diarrhoea: Cost and Cost-Effectiveness*. Geneva: World Health Organization.
4. T Clasen, I Roberts, T Rabie, W Schmidt, S Cairncross, 2006. Interventions to improve water quality for preventing diarrhoea. *Cochrane Database Syst Rev* **3**: CD004794.
5. S Richardson, M Plewa, E Wagner, R Schoeny, D Demarini, 2007. Occurrence, genotoxicity, and carcinogenicity of regulated and emerging disinfection by-products in drinking water: a review and roadmap for research. *Mutat Res* **636**:178-242.
6. E Mintz, J Bartram, P Lochery, M Wegelin, 2001. Not just a drop in the bucket: Expanding access to point-of-use water treatment systems. *Am J Public Health* **91**: 1565–1570.
7. F Shang, J Uber, L Rossman, 2008. Modeling reaction and transport of multiple species in water distribution systems. *Environ Sci Technol* **42**:808-14.

VALIDITY OF COMPOSITE SAMPLING FOR ENUMERATING *E. COLI* FROM RECREATIONAL WATERS BY MOLECULAR METHODS (QPCR)

J.L. Kinzelman[1] and M. Leittl[1,2]

[1]Health Department Laboratory, City of Racine, Washington Avenue, Racine, WI 53403 USA
2Department of Applied Molecular Biology, University of Wisconsin Parkside, Kenosha, WI 53144, USA

1 INTRODUCTION

Since 1968 the federal government of the United States has recommended routine recreational water quality monitoring using bacterial indicators[1]. Current water quality guidelines recommend, and coastal water quality regulatory agencies have adopted, the use of single sample advisory limits based on the bacterial indicators *Escherichia coli* (*E. coli*, fresh water) and enterococci (marine water)[2,3] as a means of protecting swimmer health. The BEACH Act (2000)[4], as implemented by the Wisconsin Department of Natural Resources (WDNR), requires one surface water sample to be collected every 500 m at designated swimming beaches. At large beaches, a single sample, collected at a single moment in time, may not be representative of ambient water quality. In the interest of public health, increasing spatial coverage is desirable to increase sampling reliability but it will increase the costs associated with routine recreational water monitoring programs, most significantly the cost associated with multiple laboratory analyses[5]. To expand coverage while controlling costs, beach managers contractually required or voluntarily collecting multiple samples may be candidates for composite sampling[6,7]. Compositing is the physical combining of two or more sub-samples into a single sample for laboratory analysis. Composite sampling may be useful if the act of compositing does not mask true elevations due to spatial variability and as long as individual samples can be adequately homogenized without affecting the integrity of the sample or introducing bias[8,9,10]. In a previous study (North and Zoo Beaches; Racine, WI; 2003), Kinzelman et al. (2006)[8] demonstrated that composite sampling was a valid procedure with an approximate 1:1 ratio between the composite sample and the arithmetic mean of the three (Zoo Beach) or four (North Beach) individual samples when using *E. coli* as the faecal indicator bacteria (FIB) and a culture-based assay (IDEXX Colilert-18/Quanti-Tray 2000).

Although composite sampling may increase spatial coverage without significantly altering regulatory action, the significant lag time between receipt of sample and reporting of results (18 – 24 hours) when using culture-based assays fails to accurately to characterize health risk as microbial water quality may change rapidly[11]. For this reason, the United States Environmental Protection Agency (US EPA) has been mandated to include rapid assessment methods, such as quantitative polymerase chain reaction (qPCR) in their revised water quality criteria[12]. Critical science related to the utility of qPCR for monitoring coastal was completed by December 2010 but outstanding questions remain, i.e. whether techniques, like composite sampling, will remain valid when using a molecular assay as the quantification method for faecal indicator bacteria (FIB). Therefore, the purpose of this study was to determine if composite sample analysis was still acceptable at the previous study sites (North and Zoo Beaches) when the bacterial target (*E. coli*) was quantified using qPCR. Specifically, whether or not the composite sample value would fall within the range of the individual values and if similar regulatory decisions would be made using a molecular analytical method based on the 1986 US EPA Ambient Water Quality guidelines for fresh water[2]. If successful, the use of composite sampling would continue to help manage laboratory costs while providing a more reliable estimate of FIB density due to the near real time nature of the method (receipt of sample to results in two hours).

2 METHODS

2.1 Study Site

North and Zoo Beaches are located within the limits of the City of Racine, WI, US on the south-western shore of Lake Michigan. The principal points of sample collection were located by global positioning satellite (GPS; accurate within six m) and are listed in Table 1. For the purpose of this study, sampling events took place once weekly during the 2009 (9 events) and 2010 (11 events) bathing seasons.

Table 1 *GPS locations of monitoring points within the study area.*

Site	Latitude	Longitude	Error (± meters)
North Beach (N1)	N42° 44' 23.5"	W087° 46' 43.8"	5.2
North Beach (N2)	N42° 44' 27.3"	W087° 46' 45.7"	4.9
North Beach (N3)	N42° 44' 32.7"	W087° 46' 47.9"	4.6
North Beach (N4)	N42° 44' 37.2"	W087° 46' 49.6"	4.6
Zoo Beach (Z1)	N42° 44' 48.8"	W087° 46' 51.6"	4.0
Zoo Beach (Z2)	N42° 44' 51.2"	W087° 46' 51.9"	3.7
Zoo Beach (Z3)	N42° 44' 52.6"	W087° 46' 52.0"	4.0

2.2 Sample Collection

Surface water samples were collected from equidistantly spaced transects across North (Four ~200m transects, N1 – N4) and Zoo (Three ~200 transects, Z1 – Z3) Beaches during late spring to early autumn of 2009 (n = 9) and 2010 (n = 11) (Table 1). Each sample was collected by wading to a depth of 1.0 m and taking a 500 mL sample from approximately 0.3 m below the surface of the water using sterile Whirl-Pak™ bags (Nasco, Ft. Atkinson, WI). Samples were transported to the laboratory on ice packs and maintained at 4°C until processed (typically within 2 hours of receipt). Individual (North Beach = 4; Zoo Beach = 3) and composite samples were analyzed using both culture-based (IDEXX Colilert-18)[13] and molecular methods[14].

2.3 Culture-based Method

To remain within the level of detection for the IDEXX Colilert-18 method, typically used for compliance monitoring in Wisconsin, all samples, including the composite, were processed at a 1:10 dilution. For North Beach, which has four monitoring stations, and Zoo Beach which has three, this meant using equal portions of sub-sample from each site to achieve a total volume of 10 ml for use in the final dilution process. Therefore, from each well-mixed sub-sample an appropriately sized aliquot was withdrawn [2.5 ml from each of four samples, N1 – N4, and 3.3 ml(x2)/3.4 ml (x1) from each of three samples, Z1 – Z3] and combined to form a composite sample. To reduce sampling bias, the aliquots from Zoo Beach (Z1, Z2 or Z3) were randomized as to which would receive the 3.3 or 3.4 ml volume. Regardless of the type of analyses (individual and composite) a single well-mixed portion of surface water from each location was removed from the original sample bag (within 30 seconds of vigorous agitation[15] with a sterile, serological pipette and divided between vessels designated either for individual analysis (10 ml) or composite testing (2.5 ml or 3.3/3.4 ml). The volume of the pipette was sufficient for a single aliquot to contain enough sample material for both the single and composite analyses. The final dilution of original sample remained constant throughout the experiment; i.e. a 1:10 dilution was employed in the single sample analysis as well as in the composite analysis, as described previously.

E. coli was enumerated using IDEXX Colilert - 18® (IDEXX, Inc., Westbrook, ME), a selective cultural identification method utilizing bacterial enzymatic activity and differential substrates, for the detection of *E. coli* according to previously established laboratory protocols. In brief, diluted samples were mixed with reagent and placed in a Quanti-Tray/2000 according to manufacturer's instructions[13]. Quanti-Trays were sealed using an IDEXX Quanti-Tray® sealer and placed in a 35 °C ± 0.5 ° C incubator for 18 hours. A quality control organism (*E. coli* ATCC #25922) was run once daily to validate (qualitative) test performance, i.e. a positive test reaction. Following incubation, Quanti-Tray wells were read for yellow colour indicating onitrophenyl ß-D-galactopyranoside (ONPG) hydrolysis and fluorescence, indicating 4-methyl-umbelliferyl ß-D-glucuronide (MUG) cleavage, with the aid of a UV light box (366 nm). Wells producing fluorescence in the absence of yellow colour were determined to be false readings (*E. coli* would be classified as a total coliform and therefore should be detected by this method as such, according to the manufacturer). The number of wells producing fluorescence was compared to the MPN table provided by the manufacturer to enumerate *E. coli* as MPN/100 mL.

2.4 qPCR

2.4.1 qPCR Calibration Standards, Controls, and Standard Curves[21]. Prior to sample filtration and processing, a range of calibration standards and controls were prepared. *E. coli* (ATCC 25922; American Type Tissue Culture Collection, Manassas, VA) was grown in Brain Heart Infusion Broth (Becton Dickinson, Sparks, MD). The starting concentration of the culture was determined spectrophotometrically, and by serial dilution and plating onto Brain Heart Infusion Agar (Becton Dickinson, Sparks, MD). A stock calibration standard was made by filtering diluted cultures of *E. coli* in phosphate buffered saline (PBS, Fisher Scientific) at a concentration of 10^5 cells onto a 47 mm,0.4 μm polycarbonate membrane (GE Osmonics, Minnetonka, MN), placing in a low retention snap cap tube, and storing them at -80° C (this was the Calibration Standard stock). *Lactococcus lactis* DNA (*Lactococcus* Smart Beads™, BioGx, Birmingham, AL) was used as a specimen processing control (SPC). Serial dilutions were prepared from the frozen calibrators and analyzed at least three times to produce a master standard curve. Acceptance criteria for standard curves were: 1) an amplification factor (AF) between 1.87 and 2.1 and 2) an R^2 of greater than or equal to 0.98.

2.4.2 Sample filtration for qPCR. Undiluted surface water samples (N1 – N4 and Z1 – Z3) were filtered as individual samples (100 ml total volume) or as a composite (25 ml per sub sample, N1 – N4 or 33/34 ml per sub-sample, Z1 – Z3) using a filtration manifold and vacuum pump assembly with ground glass filter funnels and 47-mm diameter, 0.45-μm pore size polycarbonate filters (GE Osmonics, Minnetonka, MN). Prior to filtering the water sample, a filter blank was processed by filtering approximately 100 mL of sterile phosphate buffered saline (PBS). Once the sample or blank had passed through the filter, the filter funnels were rinsed with approximately 20 ml of sterile PBS, which was also filtered to visible dryness. Filters were then aseptically folded in half with the sample side facing inward, and then continually folded in half until narrow enough to fit into a 2.0-mL screw-cap micro-centrifuge tube containing 0.3 ± 0.01 g acid-washed glass beads provided in the DNA extraction kit (section 2.4.3). Filters were stored at -80°C until processed.

2.4.3 DNA Extraction. Prior to extraction, the *Lactococcus* SPC was added to each filter (unknown, calibrator, or control) in order to assess sample recovery and qPCR inhibition. DNA was extracted from the fresh or frozen filters using the Zymo ZR Faecal DNA Mini-prep Kit according to manufacturer instructions[16] (product #6010, Zymo Research, Orange County, CA). The resulting supernatants (DNA extract) were transferred to sterile 1.6-ml low-retention micro-centrifuge tubes and analyzed immediately.

2.4.4 Quantification by qPCR. *E. coli* was enumerated in surface water samples using Scorpion® qPCR chemistry (BioGx CSR *E. coli* and Cepheid OmniMix® HS master mix) on the Cepheid Smart Cycler II platform (Cepheid, Inc., Sunnyvale, CA)[14,17,18,21,22]. For each assay, 20-μl aliquots of master mix were pipetted into appropriately labelled 25-μl optical tubes (Cepheid, Inc., Sunnyvale, CA), followed by 5 μl of DNA extract. Each run included a no template control (qPCR reagents plus nuclease-free water) and a filter blank (sterile PBS). A calibrator (known quantity of *E. coli* target) was included with every run and compared with historical calibration data to assess qPCR performance (section 2.4.1). *E. coli* was expressed as calibrator cell equivalents (CCE) per 100 mL from the cycle threshold (C_T) values using the comparative C_T or $\Delta\Delta C_T$ approach[19].

2.5 Statistical analysis of data

Analysis of Variation (ANOVA), regression, and correlation analyses were performed using Microsoft Excel 2007 Data Analysis Tool-Pak. Unless otherwise stated, significance was assessed at $\alpha = 0.05$. When estimations of *E. coli* were below the limit of detection, culture-based and qPCR, the result was reported as one half the reciprocal of the dilution factor, i.e. a result of less than 10, based on no detection at a 1:10 dilution, became 5.

3 RESULTS

Mean *E. coli* density did not differ significantly across beach transects (N1 – N4 and Z1 – Z3) in either year of the study (2009 and 2010) regardless of the analytical method employed (culture-based or qPCR) ($p = 0.26 – 0.98$, Table 2) [Note that each beach was treated as a separate population]. These results indicate that compositing of samples could validly be undertaken at the study sites.

Table 2 *P-values (p, α=0.05) from single-factor ANOVA of E. coli concentrations across beach transects (NB = N1-N4, ZB = Z1-Z3); 2009 - 2010.*

	IDEXX Colilert-18®		qPCR	
Year	**NB Transects (p)**	**ZB Transects (p)**	**NB Transects (p)**	**ZB Transects (p)**
2009	0.50	0.82	0.26	0.35
2010	0.65	0.98	0.70	0.36

Log-converted raw data were compared using three different interpretation scenarios, daily geometric mean (DGM), daily arithmetic mean (DAM), and composite sample analysis, with respect to the number of bathing water quality failures triggered by *E. coli* levels in excess of the USEPA single sample standard (235 MPN/100 mL). For the culture-based method (IDEXX), composite samples were correlated, to/did not differ significantly from, the arithmetic or geometric mean of the three or four individual samples (N1 – N4 and Z1 – Z3) for all years (2009 – 2010, Table 3). Composite values fell within the range of the three (Zoo Beach) or four (North Beach) individual *E. coli* values an average of 80% of the time at Zoo Beach and 90% of the time at North Beach. When composite values fell outside of the range of the three to four individual samples, they differed by no more than 6 MPN/100 mL for North Beach and 31 MPN/100 mL for Zoo Beach [data not shown]. Using 1986 US EPA guidelines for regulatory action, based on an *E. coli* exceedance standard of 235 MPN/100 mL, the composite and individual range were in full agreement (100%). No Type I or Type II errors would have resulted from compositing samples using Colilert-18® as the analytical method.

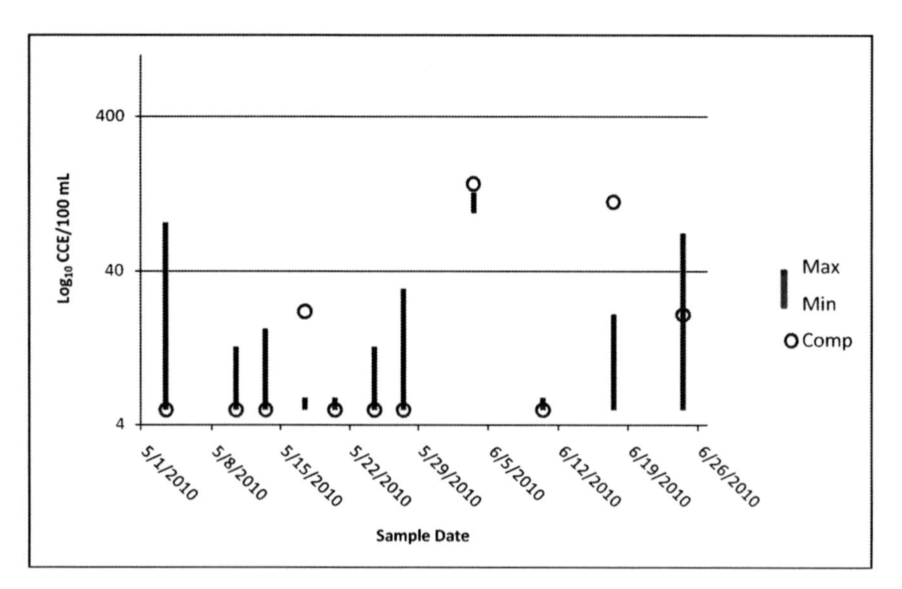

Figure 1 *North Beach E. coli composite vs. individual site analysis by qPCR (2010).*

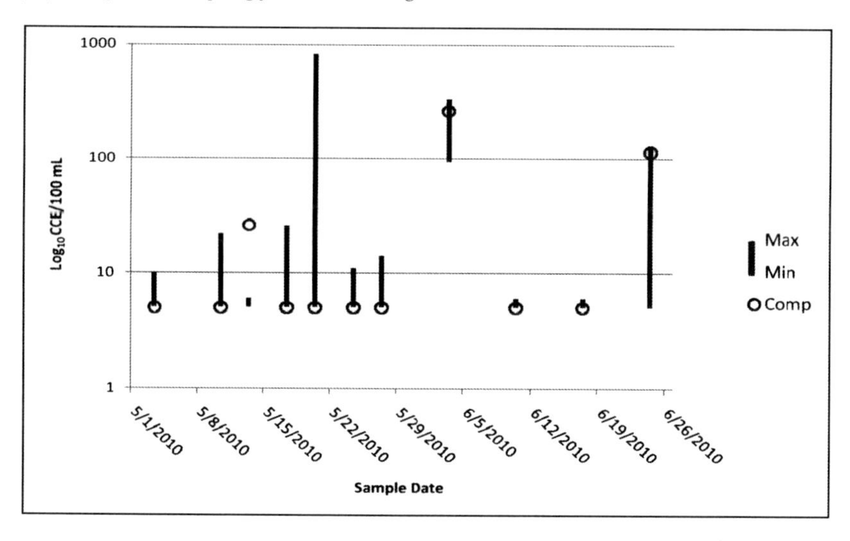

Figure 2 *Zoo Beach E. coli composite vs. individual site analysis by qPCR (2010.)*

For qPCR, composite *E. coli* samples were correlated to/did not differ significantly from the arithmetic or geometric mean of the three or four individual samples (N1 – N4 and Z1 – Z3, 2009 - 2010) (Table 3). Better correlation (R^2) was observed in 2010 than 2009. Composite *E. coli* values fell within the range of the three (Zoo Beach) or four (North Beach) individual samples an average of 85% of the time at Zoo Beach and 80% of the time at North Beach (Figures 1 and 2).

Table 3 *Relationship of E. coli concentrations in composite vs. daily geometric and daily arithmetic mean for individual beach transects (culture-based and qPCR, 2009 – 2010, α = 0.05).*

		Composite vs. Daily Geometric Mean				Composite vs. Daily Arithmetic Mean			
		p - Value		R^2		p - Value		R^2	
	Year	Colilert-18®	QPCR	Colilert-18®	QPCR	Colilert-18®	QPCR	Colilert-18®	QPCR
North Beach	2009	0.74	0.37	0.96	0.28	0.99	0.78	0.97	0.38
	2010	0.97	0.49	0.98	0.75	0.83	0.61	0.98	0.72
Zoo Beach	2009	0.77	0.39	0.99	0.79	0.84	0.76	0.99	0.18
	2010	0.86	0.51	0.95	0.91	0.91	0.76	0.94	0.41

When the composite value fell outside of the range of the three (Zoo Beach) of four (North Beach) individual values, it did so by no more than 91 CCE/100 mL at North Beach (range = 16 – 91 CCE/100 ml) and 63 CCE/100 ml at Zoo Beach (range = 14 – 63 CCE/100 ml). In every instance but one, when *E. coli* levels exceeded US EPA guidelines the composite value fell within the range of the three to four single sample values and was equivalent to their arithmetic and geometric mean values (Zoo Beach, 20 May 2010; Figure 3).

Using 1986 US EPA[2] guidelines for regulatory action based on an *E. coli* standard, a Type II error would have resulted if using composite sampling with qPCR as the analytical method on that day.

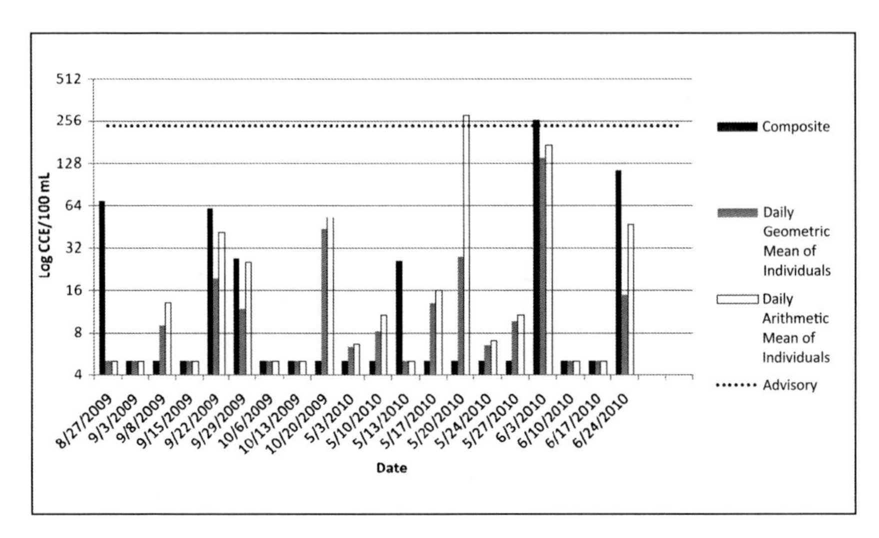

Figure 3 *Zoo Beach composite vs. daily geometric mean and daily arithmetic mean of individual samples by qPCR.*

There were four instances when the choice of analytical method (culture vs. qPCR) would have resulted in a different regulatory action. In these instances composite sample values, as determined by qPCR, did not agree with the culture-based comparable method. These disparities would have resulted in two Type II errors at North Beach (29 September 2009, 3 June 2010), one Type I error (20 May 2010) and one Type II error (29 September 2009) at Zoo Beach [Tables 4 and 5].

Table 4 *Instances when* at *North Beach a disparity existed between the composite sample as determined by qPCR and IDEXX Colilert-18® that would have resulted in an alternate regulatory action.*

North Beach									
Date	**Method**	**N1**	**N2**	**N3**	**N4**	**N-Comp**	**N-DAM**	**N-DGM**	**Regulatory Action**
09/29/09	Colilert-18® (MPN/100 mL)	583	173	121	98	259	244	186	Advisory
09/29/09	qPCR (CCE/100 mL)	128	47	5	19	28	50	27	None (Type II Error)
06/03/10	Colilert-18® (MPN/100 mL)	189	181	318	457	238	286	266	Advisory
06/03/10	qPCR (CCE/100 mL)	95	98	130	109	148	108	107	None (Type II Error)

On the majority of occasions when the single sample results were unequivocal in terms of whether the single-sample limit had been exceeded, (North Beach = 3 - 4 values above or below 235 MPN/100 ml, Zoo Beach = 2 - 3 values above or below 235 MPN/100 ml), the composite sample result was in agreement. Discrepancies with regard to single-sample limit exceedance were due to the value obtained at a single sampling point (Tables 4 and 5). This holds true for the entire study, with only one exception, North Beach, 3 June 2010 (Table 4). It could be argued that, on such occasions, the composite sampling process is masking a result that indicates a health risk to bathers. However, on only one of these occasions did more than one of the three or four individual samples exceed the threshold value, illustrating the "hit-or-miss" nature of reliance upon single sample results.

Table 5 *Instances at Zoo Beach when a disparity existed between the composite sample as determined by qPCR and IDEXX Colilert-18® that would have resulted in an alternate regulatory action.*

Zoo Beach								
Date	**Method**	**Z1**	**Z2**	**Z3**	**Z-Comp**	**Z-DAM**	**Z-DGM**	**Regulatory Action**
09/29/09	Colilert-18® (MPN/100 mL)	359	134	146	295	213	192	**Advisory**
09/29/09	QPCR (CCE/100 mL)	67	5	5	27	26	12	**Open Type II Error**
05/20/10	Colilert-18® (MPN/100 mL)	5	5	5	5	5	5	**Open**
05/20/10	QPCR (CCE/100 mL)	5	5	828	5	279	27	**Advisory Type I Error**

4 CONCLUSIONS

Composite sampling increases spatial coverage and sampling reliability without increasing analytical costs. Currently approved culture - based methods typically take 18-24 hours to generate a result, during which time water quality may have changed significantly. Rapid molecular methods, such as qPCR, are appealing due to their ability to generate results in near real-time, hence providing better protection of public health. The cost differential between qPCR and culture-based methods may be significant[20]. Therefore, composite sampling may become even more attractive to those beach managers opting to implement rapid analytical methods for the purpose of regulatory monitoring.

The lack of significant difference across beach transects (N1 – N4 and Z1 – Z3) (both methods/both years) indicate that composite sampling remains a viable option for beach management in Racine, WI where compliance with proposed revisions to the BEACH Act will likely require use of a rapid analytical method. A lack of consistent correlation, as observed to some degree in 2009, may be due to an underestimation bias for qPCR potentially attributed to different types of contamination sources (age and growth state of faecal indicator bacteria), unresolved inhibition, method variability (proficiency of analyst with the assay, extraction efficiency, and limit of detection), and the relative contribution of DNA from viable vs. non-viable cells[20, 21].

Molecular methods, such as qPCR, may not be unilaterally applicable. Prior to implementing molecular assays, with or without composite sampling, analysts should be thoroughly trained in the method and how to recognize and compensate for possible obstacles, such as inhibition, leading to underestimation of target. If implemented, a periodic analysis of individual samples may be needed for ongoing validity purposes.

Composite sampling can save time during processing and/or DNA extraction, better characterize water quality by increasing the sampling reliability, decrease analytical costs and potentially reduce Type I and Type II errors (through the generation of near real time results). In this study, qPCR was successfully used to enumerate *E. coli* via individual sample analysis with averaging and composite sampling as a rapid alternative to culture-based assays without significant impact on regulatory action.

References

1 National Technical Advisory Committee. Department of the Interior, Washington, D.C., 1968.
2 *Ambient Water Quality Criteria for Bacteria.* USEPA, United States Office of Water, 1986.
3 *Improved Enumeration Methods for the Recreational Water Quality Indicators: Enterococci and Escherichia coli.* USEPA, Office of Science and Technology, 2000.
4 *Beaches Environmental Assessment and Coastal Health Act of 2000.* United States public law 106-284.
5 R.L. Whitman and M. B. Nevers, *Environ. Sci. & Technol.*, 2004, **38(16)**, 4241-4246.
6 E.E. Bertke, E.E., *J. Great Lakes Research*, 2007, 33(2):335-341.
7 *Sampling and Consideration of Variability (Temporal and Spatial) For Monitoring of Recreational Waters, EPA-823-R-10-005.* US EPA, United States Office of Water 2010.
8 J. Kinzelman, A. Dufour, L. Wymer, G. Rees, K. Pond, and R. Bagley, *Lake and Reservoir Manag.*, 2006, **22 (2):** 95-102
9 A.H. El-Shaawari and S.R Esterby, 'Sampling Recreational Waters' in *Statistical Framework for Recreational Quality Criteria and Modeling*, John Wiley and Sons, West Sussex, U.K, 2007, pp 69 - 91.
10 J.D. Reicherts and C.W. Emerson. 2009. *Environ. Monit. Assess.*, 2010 Sep; **168 (1-4)**: 33-43.
11 A.B. Boehm, N.J. Ashbolt, J.M. Colford Jr., L.E. Dunbar, L.E. Fleming, M.A. Gold, et al., *J Water Health*, 2009, **7(1)**:9–20.
12 United States District Court for the Central District of California, Case 2:06-cv-04843-PSG-JTL Document 160 Filed 09/04/2008.
13 Colilert-18® package insert. IDEXX Laboratories, Inc. Westbrook ME.
14 R.R. Converse, J.F. Griffith and R. T. Noble, *J. Wat. Res.*, 2009, **43(19)**:4828-37.
15 *Standard Methods for the Examination of Water and Wastewater*, 20th edition. Clesceri, L., Greenburg, A., and Eaton, A. (eds.), American Public Health Association, 1998.
16 Zymo ZR Faecal DNA Miniprep Kit package insert (product 6010), Zymo Research, Orange County, CA.
17 J.F. Griffith and S.B. Weisburg, *Southern California Coastal Water Research Project*, 2007.

18 J. Lavender and J. Kinzelman, 2009. *Wat. Res.*, 2009, **43(19)**, 4967 – 4979.
19 R. Haugland, S. Siegfring, L. Wymer, K. Brenner, and A. Dufour. 2005. *Wat. Res.*, 2005: 559 – 568.
20 J.L. Kinzelman, unpublished results.
21 R.T. Noble, A.E. Blackwood, J.F. Griffith, C.D. McGee and S.B. Weisberg, *Appl. Environ. Microbiol.*, 2010, **76(22)**: 7437-7443.
22 J.L. Kinzelman, R.N. Bushon, S. Dorevitch, and R.T. Noble, WERF Report PATH7R09. 30 June 2011.

ESTIMATING 95th PERCENTILES FROM MICROBIAL SAMPLING: A NOVEL APPROACH TO STANDARDISING THEIR APPLICATION TO RECREATIONAL WATERS

R.S.W. Lugg,[1,2] A. Cook,[2] B. Devine[2]

[1] Department of Health, Western Australia, East Perth, WA 6004, Australia
[2] School of Population Health, The University of Western Australia, Crawley WA 6009, Australia

1 INTRODUCTION

The recreational water guidelines issued by the World Health Organization (WHO)[1] and similar risk-based guidelines such as those from Australia[2] and New Zealand,[3] are concerned to estimate the excess risk of gastrointestinal illness following whole body immersion, relative to non-bathers (*infection risk*). They use an equation developed by Wyer *et al.*[4] that derives this infection risk from the distribution of the faecal indicator *enterococci* in recreational waters. These guidelines assess the microbial quality of recreational waters from the 95th percentiles of *reference distributions* of enterococci, that is, lognormal distributions with a $\log_{10}$ standard deviation (SD) of 0.8103. A subset of these reference distributions (namely, those with 95th percentiles of 40, 200 and 500 enterococci per 100mL) have been chosen as the boundary distributions demarcating microbial water quality assessment categories (MACs) A, B and C respectively, each with a defined range of infection risk.

A number of methods of calculating 95th percentiles of enterococcal sampling results are in common use for the purpose of assessing the microbial quality of recreational waters in terms of the above MACs. However all are liable to give a poor indication of infection risk as estimated by the Wyer *et al.*[4] equation, unless the distribution of the sampling results approximates a reference distribution as defined above. Where the distribution of sampling results varies significantly from a reference distribution, there is a substantial risk that the water will be placed in the wrong MAC: that is, one in which the implicit infection risk range does not include the actual infection risk of the water (as estimated by the above equation). This issue of misclassification is a major focus of this paper.

There is a demand among recreational water managers for faecal indicator trigger levels that would facilitate a prompt management response to any real or apparent deterioration in water quality suggested by routine sampling. Most risk-based guidelines do not recommend any trigger levels,[1,2] or recommend only generic levels that may be too low to be of much practical application.[3]

1.1 The Wyer *et al.* equation

The present paper takes this equation as a given, and seeks to facilitate the application of recreational water guidelines that use it for the purpose of allocating MACs to recreational waters. There are a number of criticisms of the equation – such as the objection that it may overestimate infection risk in freshwaters or underestimate the risk for young children[1,2] – but such criticisms will not be examined further here.

The equation is:

$$y_{\mathrm{i}} = 1/(1+e^{-\mathrm{m}}) - 0{\cdot}0866 \tag{1}$$

where y_{i} = infection risk at enterococcal concentration of x_{i} per 100mL

$$\mathrm{m} = 0{\cdot}20102(x_{\mathrm{i}} - 32)^{0.5} - 2{\cdot}3561 \text{ for i} = 33\text{-}158$$

$$y_{\mathrm{i}} = y_{158} \text{ for i} > 158.$$

Two peculiarities of equation (1) should be noted:

- It assumes a threshold value of 33 enterococci per 100mL, below which the infection risk does not exceed that for non-bathers; and
- It specifies a plateau at 158 enterococci per 100mL, above which the infection risk is presumed not to rise.

The outputs for equation (1) are shown graphically in Figure 1.

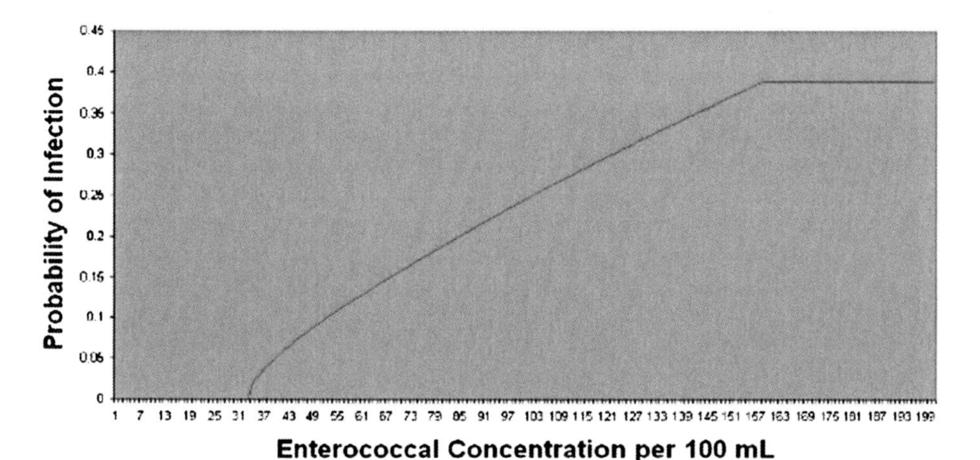

Figure 1 *Infection risk in recreational waters, according to equation (1).*

1.2 The problem of misclassification

1.2.1 Using 95th percentiles from lognormal distributions. Provided the distribution of any set of enterococcal sampling results is known, the infection risk may be estimated by applying equation (1). Where the distribution is lognormal, the infection risk will vary with the two defining parameters of a lognormal distribution, its geometric mean and its log SD. Both these parameters are reflected in the 95th percentile, making it an attractive summary statistic for the distribution. However, this statistic will not correlate well with the MACs as defined in the guidelines mentioned above, unless the distribution approximates a reference distribution, i.e. a lognormal distribution with a $\log_{10}$ SD of about 0.81. The effect of varying the log SD is shown in Table 1.

Table 1 *Exact boundary values (expressed as 95th percentiles) between the various MACs, for lognormal enterococcal distributions carrying the same infection risks as those in guidelines (in which the $\log_{10}$ SD is held constant at 0.8103).*

$\mathbf{Log_{10}}$ SD	**0.4**	**0.5**	**0.6**	**0.7**	**0.8**	**0.9**	**1.0**	**1.1**
MAC A	46	45	43	42	40	38	36	34
MAC B	110	128	149	171	197	226	258	297
MAC C	181	233	299	382	488	621	791	1006

The figures in the sixth column are very close to the boundary values in the guidelines (40, 200 and 500), but the further away the log SDs deviate from 0.81, the greater the opportunity for misclassification. The effect is minimal for MAC A, and running counter to the other MACs, because of the threshold effect, in which levels below 33 do not contribute to infection risk. This means the tighter distributions (with a low log SD) must have a considerably higher geometric mean, pushing up the 95th percentile in comparison with the more scattered distributions (with higher log SDs).

1.2.2 Using 95th percentiles where the distribution is not lognormal, or is not known. Using a conventional *parametric 95th percentile* (calculated from the geometric mean and log SD) where the distribution is either known to be other than lognormal, or is not known at all, does not permit conclusions to be drawn with any confidence as to the infection risk carried by that distribution. As a result, it is uncertain which is the correct MAC to which the water should be assigned.[5] Unfortunately, despite strong advocacy for certain *non-parametric methods* of calculating the 95th percentile,[1,5] the use of any non-parametric method is no more informative about the distribution, its associated infection risk, or the correct MAC to which the water should be assigned. Hence none of these methods can be recommended to recreational water managers tasked with applying any of the risk-based guidelines for the proper management of the infection risk to bathers. [A non-parametric approach to the setting of trigger levels is described in section 3.2.]

1.3 The concept of standardisation

As stated above, if the distribution of any set of enterococcal sampling results is known, the infection risk may be estimated by applying equation (1). Once the infection risk is known, that distribution can be matched with the reference distribution that carries the same infection risk. The 95th percentile of that reference distribution then becomes the *standardised 95th percentile* of the original distribution. As with all applications of the principle of standardisation, the intent is to allow direct comparison of different sets of data, in this case in terms of their infection risk. It also allows the water to be placed in its correct MAC (namely the one whose underlying infection risk range encompasses the infection risk of the data set in question).

2 OBJECTIVES

The main aim of this study was to develop and test an automated, easy-to-use instrument for standardising 95th percentiles of enterococci in recreational waters, thereby accurately reflecting the infection risk as calculated by equation (1). It was necessary for the instrument to be able to estimate infection risk, no matter how the sampling results for the enterococci were distributed. It was also necessary for the instrument to be able to accommodate the common problem of *left-censoring*, where the laboratory sets a minimum enumerated value below which results are reported as being below the limit of detection.

A secondary aim was to provide trigger levels that – when exceeded – would signal possible deleterious changes in the underlying distribution, warranting a management response such as additional sampling, or investigation of possible changes in the sanitary conditions affecting the recreational water body in question.

A third aim was to explore the degree of uncertainty associated with the process of standardising 95th percentiles,

3 METHOD

A Microsoft® Excel template (the *EnteroTester*) has been developed for generating workbooks that automate and record the calculation of standardised enterococcal 95th percentiles for data sets containing eight observations or more. The template comes in two versions: a full version, which can handle up to 677 observations in a data set (although it could be readily extended to handle more), and an equally functional lighter version intended for regular use, which handles up to 200 observations. These versions are copyrighted by the State of Western Australia, but are freely available without charge for non-commercial purposes as zip files from the State Department of Health. The lighter version (intended for regular use) is available from www.public.health.wa.gov.au/3/662/2/bacterial_water_quality.pm (go to the link template "(the Enterotester)"). Both versions are available from www.public.health.wa.gov.au/3/1287/2/publications.pm (go to the heading "Forms and Templates"). In addition to the templates, the zip files contain simple step-by-step instructions for use, and a security certificate designed to streamline the running of the macros associated with the templates.

3.1 Procedure for standardising 95th percentiles

3.1.1 Testing for lognormality. The data are ranked and log-transformed, and the transformed data paired with their corresponding Blom normal order scores from the reference distribution with a 95th percentile of 200 enterococcci per 100mL, for computing the Shapiro-Francia statistic W', which is then tested for significance by the method of Royston.[6] The data are first prepared by replacing any zeroes with the left-censored value <1; ties are dealt with by taking the mean value of the Blom normal order scores for each observation in a tie. This method permits up to 80% of the data to be left-censored, provided there are at least 20 observations in the data set.[7]

3.1.2 Proceeding to standardisation. Once the data have been prepared for further analysis, the EnteroTester proceeds by default to standardise the 95th percentile by calculating the infection risk of the distribution. However, there is an option to assign a 95th percentile directly, and this option is offered if the lognormal assumption is probable ($p > 0.5$) and the sample log SD is close to 0.81 (differing from it by < 1% of the standard error of the reference log SD, a parameter that decreases as the number of observations increases). This *assigned 95th percentile* is the above reference 95th percentile (200 enterococcci per 100mL), multiplied by the ratio of the geometric means of the observations and their corresponding Blom normal order scores (from the above reference distribution) for the top half of the distribution.

3.1.3 Accepting the lognormal assumption. The lognormal distribution is taken as the default distribution for enterococci in recreational waters,[8] and the EnteroTester will recommend accepting the lognormal assumption so long as the probability of lognormality is 5% or more. However the EnteroTester does allow the user to compare the effects of accepting and rejecting the lognormal assumption, if desired.

3.1.4 Rejecting the lognormal assumption. If the probability of lognormality is less than 5%, the EnteroTester will recommend rejecting the lognormal assumption, in which case the infection risk will be calculated from the empirical distribution of the data set, i.e. the actual sampling results. This carries more uncertainty than when the lognormal distribution is used, although the difference is not great (see section 4.1.1).

3.1.5 Calculating the infection risk.. The Enterotester template incorporates a look-up table of infection risks for each integral number of enterococci per 100mL, from 32 to 158 (as shown graphically in Figure 1). When the empirical distribution is used, the infection risk of the distribution is calculated by summing the risks for every observation in the data set and dividing by the total number of observations. When the lognormal assumption is applied, the weighting for every number in the table is calculated parametrically, and the proportionate risks for each of them summed to give the overall infection risk of the distribution.

3.1.6 Transforming to 95th percentiles. The 95th percentile of the reference distribution with the same infection risk as that arrived at for the distribution is back-calculated using one of a series of four quartic polynomial regression equations, developed for the template. The equation is:

$$0.005754568r^4 - 0.069483077r^3 + 2.215781121r^2 + 24.42801421r + 10.61697665,$$

where r is depending on the range into which the infection risk of the distribution. The result is then rounded to approximately two significant figures to give the *standardised 95th percentile* of the original distribution, and its MAC is allocated accordingly.

3.2 Setting trigger levels

For each data set analysed, the EnteroTester provides trigger levels intended to signal a possible deleterious change in the underlying distribution. Since an appropriate management response is presumed, it is considered undesirable that this signal should be given too frequently under normal (essentially unchanged) circumstances, lest it should have the effect of inducing complacency. The *false trigger rate* (rate that would be exceeded by "normal" results, lying in the right tail of a lognormal distribution) has been set at 2%, that is, one that would be exceeded only once in two to three years if samples are collected at a rate of 20 per year. The false trigger rate is comprised of two roughly equally likely scenarios: a result exceeding the 99th percentile of the distribution (*one-off trigger level*); or two consecutive results exceeding the 90th percentile of the distribution (*two-in-a-row trigger level*).

If the EnteroTester recommends accepting the lognormal assumption ($p \geq 0.05$), these trigger levels are estimated parametrically; otherwise, a non-parametric approach is adopted, using the average of the Hazen and Blom methods, as described in Appendix 3 of the NHMRC guidelines.[2] For the reasons given by Hunter,[9] interpolations between data points are carried out logarithmically. At least 57 observations are needed to calculate the 99th percentile by this method; if there are fewer observations, the trigger level defaults to an approximation of the 99th percentile of the reference distribution forming the upper bound to the MAC in which the distribution has been placed. If the distribution has been placed in MAC D (standardised 95th percentile > 500 enterococci per 100mL), it returns the expression "Nil available." The 90th percentile, on the other hand, is always available.

4 RESULTS

The performance of the EnteroTester has been assessed through various Monte Carlo simulations, and also through practical application at coastal beaches and freshwater sites in several jurisdictions in Australia.

4.1 Monte Carlo simulations

Risk-based guidelines that use equation (1) have not typically been concerned with estimating the uncertainty associated with calculating 95th percentiles, although the 2009 Addendum to the WHO guidelines[1] does give rates of misclassification for parametric 95th percentiles calculated from different numbers of samples in a data set. The statistical basis for calculating the uncertainty (scatter) associated with estimating parametric 95th percentiles has been given for normal distributions by McBride.[10] This work allows the reduction in uncertainty with increasing numbers of samples to be quantified, and provides a benchmark against which to measure the uncertainty associated with standardising 95th percentiles by the EnteroTester method.

Modified templates were developed that enabled repeated Monte Carlo simulations of the EnteroTester method to be run on randomly generated data sets from the reference lognormal distribution having a parametric 95th percentile of 200 and a $\log_{10}$ SD of 0.81. Simulations were run for different sample sizes (from 8 to 100) and different proportions of standardisations using the empirical (that is, the random) distribution (from 0% to 100%). All values below 10 (amounting to about 51.5% of the total distribution) were treated as censored. The opportunity was taken to compare the uncertainty associated with the Blom, Hazen, Excel and arithmetic (that is, percentage compliance) methods of calculating 95th percentiles non-parametrically from the ordered data sets.

4.1.1 Dispersion (SDs). As expected, the non-parametric methods all fared worse in terms of scatter (uncertainty) than the parametric and Enterotester method (see Table2). The Enterotester's standardisation approach performed better than any of the others, varying from 23.8% in 5000 simulations in which the lognormal assumption was exclusively applied, to 26% in 5000 simulations where only the empirical distribution was used. In 10,000 simulations mirroring the template in rejecting the lognormal assumption where its probability was less than 5%, the corresponding figure was 24.5%.

Table 2 *Measures of dispersion and central tendency amongst various 95th percentiles, for simulations of 100 data points each.*

Type of 95th percentile	Number of simulations	SD as % of 200 cfu/100mL	Mean Values	Difference from expected value (%)
Blom	31,000	45.6	217.9	+8.0
Hazen	31,000	44.2	213.3	+5.7
Excel	6,000	40.6	197.2	–2.3
Arithmetic	6,000	40.3	195.2	–3.3
Parametric	70,000	34.8	205.0	+1.6
Standardised	70,000	24.8	197.5	–2.1
Assigned	6,000		208.5	+3.3

4.1.2 Central tendency (mean values). Due to an asymmetry in one-sided confidence intervals about percentiles above the median,[10] the 50% upper confidence bound for the 95th percentile of the distribution used in these Monte Carlo simulations turns out to be 202, rather than the nominal value of 200. Again, the non-parametric methods fared worse than the parametric and EntroTester methods in terms of distance from this expected value, with the Blom and Hazen methods being above and the Excel and arithmetic methods below (see Table 2). Where the lognormal assumption was exclusively applied, the Enterotester gave a mean value of 196.7 (2.5% below the level), but where only the empirical distribution was used, the mean value was 199.1 (1.4% below). Where the template criterion for rejecting the lognormal assumption was mirrored, the mean value was 197.5 (2.2% below the level). Interestingly, when the standardised 95th percentile was divided by the Shapiro-Francia statistic *W'*, a much closer result was obtained, averaging 202.3 (0.2% above the level, with a range of 201.2 to 203.9 (0.3% below to 1.0% above the level). Mirroring the template criterion returned a mean of 202.2 (0.2% above the level). The rarely used *assigned 95th percentile* (see section 3.1.2) returned a mean value of 208.5 (3.3% above the expected level).

4.1.3 Upper confidence limits. Other simulations were carried out in which fewer data points were used, causing an increase in the size of the SDs, and hence of the uncertainty, or confidence intervals. 5000 simulations each were run for a series of different sample sizes, from 10 to 80. The effect is quite marked (see Figure 2). This Figure also shows the effect of the size of the sample on the highest standardised 95th percentile for which the 95% upper confidence limit does not exceed 200. In all cases, the standardised 95th percentiles track substantially above the corresponding parametric estimates calculated by the method of McBride,[9] This means that by standardising 95th percentiles, fewer data points are required to provide the same degree of confidence as would be obtained from parametric calculations. For instance, from Figure 2 it can be seen that a standardised 95th percentile of 110 enterococci per 100mL, based on 30 observations, provides the same degree of confidence that the water should be classified in MAC B as the same 95th percentile arrived at parametrically from 80 observations.

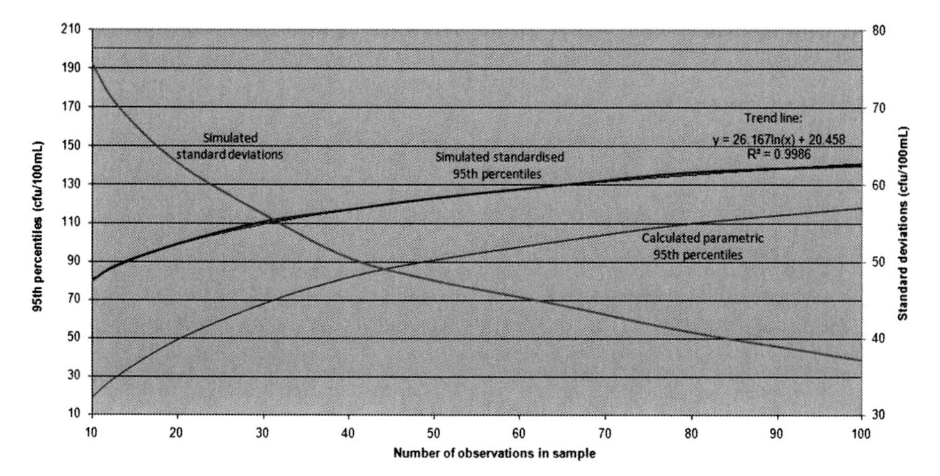

Figure 2 *Simulated standard deviations and highest standardised and parametric 95th percentiles for which there is 95% confidence that the underlying (true) value does not exceed 200 enterococci per 100mL.*

Recreational water managers whose sampling resources may be limited can thus benefit by using standardised 95th percentiles. Depending on results, it may be possible to reduce drastically the number of samples required to be confident of the MAC in which a recreational water body should be placed. Table 3, which is based on the trend line seen in Figure 2, shows that in many situations it may not be necessary to take more than 25 samples, as in the example of taking five per year over the five-year period usually allowed by risk-based guidelines.[1,2,3]

Table 3 *Highest standardised 95th percentiles for which there is 95% confidence that the underlying (true) value does not exceed the level shown.*

Size (n) of data set	40 cfu/100mL	200 cfu/100mL	500 cfu/100mL
10	16	81	202
15	18	91	228
20	20	99	247
25	21	105	262
35	23	113	284
50	25	123	307
65	26	130	324
80	27	135	338
100	28	141	352

4.2 Practical applications

The EnteroTester has been trialled for several years at coastal beaches and freshwater sites within the metropolitan area of Perth, Western Australia, and more recently at other sites in Western Australia and in New South Wales and the Northern Territory of Australia. An evaluation of its role in the assignment of provisional risk classifications for 27 sites in the Swan and Canning Rivers in the Perth metropolitan area has shown it to be an effective tool in determining MACs.[11] Similar data are available for 37 coastal beaches in the metropolitan area,[12] and for 17 coastal beaches and 10 freshwater sites in regional Western Australia.[13,14]

A total of 256 swimming locations along the New South Wales coast are monitored under the Beachwatch program, including 127 sites in the Sydney, Hunter and Illawarra Regions, and a further 129 sites are monitored in partnership with local councils.[15] The EnteroTester has been successfully used as part of the Beach Suitability Grade assessment for annual State of the Beaches reports.[16] The tool was also employed in a study of 17 beaches in the Hunter Region, in which the contribution to infection risk from rainfall exceeding 5mm in the 48 hours before sampling was removed, in order to better pinpoint sites being impacted by a dry weather pollution source.[17] It has also been applied to adjust site-specific rainfall thresholds used to predict the likelihood of pollution for real-time pollution advisories at Sydney's beaches.[18]

Users have found the automated format of the EnteroTester easy to apply and interpret, providing a fast and efficient means of calculating MACs for a large number of sites. A particular convenience is the way in which analysed data may be kept on a spreadsheet, and added to for recalculation at later intervals.[19]

The study of sites in the Swan and Canning Rivers in the Perth metropolitan area found that trigger levels produced by the EnteroTester helped authorities respond to an unanticipated deterioration in water quality that was atypical for a site, rather than relying on generic trigger levels that may not be indicative of a potential health risk.[11] This is particularly true where enterococcal levels are normally in the mid-range of acceptable values. For very good quality recreational water, the trigger levels may be too low to warrant a specific management response, while for poor quality waters, management may need to respond well short of the very high trigger levels generated by the EnteroTester. This is an area that requires further investigation.

5 CONCLUSION

The EnteroTester template is rapid, convenient, reliable, efficient in terms of data requirements, and more accurate in estimating infection risk and in placing recreational waters in their correct MACs than other methods of calculating 95th percentiles. It is available on the web from the State Department of Health in Western Australia without charge for non-commercial use, and should be considered for recreational water classification and management in accordance with risk-based guidelines such as those published by WHO.

References

1 *Guidelines for safe recreational water environments*, WHO, 2003; Addendum, 2009.
2 *Guidelines for Managing Risks in Recreational Water*, NHMRC [Australia], 2008.
3 *Microbial Water Quality Guidelines for Marine and Freshwater Recreational Areas*, New Zealand Ministry for the Environment and Ministry of Health, 2003.
4 M.D. Wyer, D. Kay, J.M. Fleisher, R.L. Salmon, F. Jones, A.F. Godfree, G.Jackson and A.Rogers, *Water Research*, 1999, **33**, 715.
5 R. Chawla and P.R. Hunter, *Water Research*, 2005, **39**, 4552.
6 P. Royston, *The Statistician*, 1993, **42**, 37.
7 S. Verrill and A. Johnson, *J. Amer. Statistical Assoc.*, 1988, **83**, 1192
8 J. Bartram and G. Rees, eds., *Monitoring Bathing Waters*, E and FN Spon, London, 2000.
9 P.R. Hunter, *Letters in Appl. Microbiology,* 2002, **34**, 283.
10 G. McBride, *Using Statistical Methods for Water Quality Management*, 2005
11 B. Abbott, R. Lugg, B. Devine, A. Cook and P. Weinstein, *J. Water Health*, 2011, **9**, 70.
12 *Metropolitan Ocean Bacterial Water Quality Monitoring Program: Provisional Risk Classification for 2009/2010*, Department of Health, Western Australia, 2010 <www.public.health.wa.gov.au/cproot/3424/2/0b01KJ_LG_MO_RiskClass.pdf>, accessed 10 May 2011.
13 *Regional Recreational Waters Bacterial Monitoring Program: Provisional Risk Classification for 2009/2010*, Department of Health, Western Australia, 2010 <www.public.health.wa.gov.au/cproot/3426/2/0b01KJ_LG_Reg_RiskClass.pdf>, accessed 10 May 2011.
14 *Rottnest Island Bacterial Ocean Water Quality Sample Results 2003-2009*, Department of Health, Western Australia, 2010 <http://www.public.health.wa.gov.au/cproot/3368/2/Rottnest%20Island%20Bacterial%20Ocean%20Water%20Quality%20Sample%20Results%202004-2009.pdf>, accessed 10 May 2011.
15 *Beachwatch Programs*, Office of Environment & Heritage, Department of Premier and Cabinet, New South Wales, 2011 <www.environment.nsw.gov.au/beachapp/default.aspx>, accessed 10 May 2011.
16 Beachwatch Programs, Office of Environment & Heritage, Department of Premier and Cabinet, New South Wales, 2010 <www.environment.nsw.gov.au/beach/reportann.htm>, accessed 31 May 2011.
17 C. Hickey, unpublished study.
18 C. Hickey, personal communication.
19 B. Abbott, personal communication.

COMPARISON OF RAPID METHODS FOR ACTIVE BATHING WATER QUALITY MONITORING

A. Henry[1], G. Scherpereel[1], R.S. Brown[2], J. Baudart[3], P. Servais[4], N. Charni Ben Tabassi[1].

[1]Veolia Environnement Recherche et Innovation, 1, place de Turenne, 94417 Saint Maurice, France. E-mail: annabelle.henry@veolia.com
[2]ENDETEC, 116 Barrie St., Kingston, K7L 3N6, Canada
[3]UPMC Université Paris 06, CNRS, UMR 7621, LOMIC, Observatoire Océanologique, F-66650 Banyuls/mer, France
[4]ULB, Campus de la Plaine, Boulevard du Triomphe, 1050 Bruxelles, Belgique

1 INTRODUCTION

In France, recreational water quality is at the junction of three categories of issues: medical, environmental, and economic. In the seventies, the European policy on water was established to limit the pollution of bathing sites in order to protect bathers from potential health risks.[1] This Directive considers the obligation for member states to follow the water quality of identified or designated bathing sites. A new European Directive for the management of bathing water quality was published in 2006.[2] It brings a legal framework to these problems and increases the responsibility of coastal towns in the monitoring of the sanitary risk of their bathing sites. This Directive also introduces the concept of bathing water active management and suggests the implementation of the following technical actions:(i) assess contamination vulnerability of the bathing area; and (ii) perform regular analyses of water for both indicators of faecal contamination i.e. *Escherichia coli* (*E. coli)* and intestinal enterococci.

The species *E. coli,* which belongs to the group of faecal coliforms, is found at high concentrations in the faeces of mammals, and is generally considered as not growing and reproducing in the aquatic environment. *E. coli* is recognized by US health departments as one of the best indicators of faecal pollution and the presence of potential pathogenic microorganisms.[3,4]

The detection of *E. coli* is commonly used in France as the current standard instead of faecal coliforms.[2] *E. coli* shows a specific enzymatic activity which is useful for its identification and enumeration.[5,6,7] Thus, approximately 97% of *E. coli* present will express the β-D-glucuronidase enzyme, in contrast to the majority of other coliform bacteria.[8] In Europe, the reference method usually used for *E. coli* enumeration in bathing waters is the "miniaturized method by inoculation in liquid medium". It is based on glucuronidase enzyme detection and is described by a standard procedure for surface water and waste water.[9] During the incubation (from 36 to 72 hours at 44°C), the target bacteria in the sample multiply and hydrolyze the substrate with a release of a fluorescent blue compound. This miniaturized method estimates the quantity of *E. coli* by the Most Probable Number (MPN) approach. Consequently, this method can only detect the viable and culturable bacteria (VC) and cannot detect the viable but not cultivable cells (VBNC) or the dead cells.[10] It may also detect non *E. coli* bacteria which express β-D-glucuronidase.[11,12,13] The time needed to obtain a response (36-72h) with this method is a major disadvantage within the 2006 Directive cited above which requires real time detection. Indeed, Noble and Weisberg defined a "rapid" method as one which provides a result in "less than four hours".[14] Thus, several rapid assays have been developed for enumerating *E. coli* such as molecular, impedance and indicator enzyme based methods.

Molecular-based methods include immunological methods, polymerase chain reaction (PCR) and *in-situ* hybridization (ISH) techniques.[15,16,17] These techniques are not easily applicable for the routine monitoring of bathing water quality due to their complex implementation, the need of an equipped laboratory with an experienced technician and cost. Moreover, ISH without a direct viable count step cannot discriminate between VC and VBNC cells.

Culture coupled with impedance or indicator enzyme-based microbiological methods are commonly used to detect *E. coli* in live shellfish or drinking water. [18,19,20] BioRad produces the XplOrer technology based on an impedance measurement.[21] ENDETEC provides a technology based on an indicator enzyme measurement. [19] Each ENDETEC cartridge contains pre-measured amounts of growth media that support the enrichment of any target bacteria that are present in the sample. As target bacteria such as *E. coli* or total coliforms begin to multiply, they emit a specific enzyme that interacts with a proprietary chemical substrate in the cartridge, releasing fluorescent molecules. These fluorescent molecules rapidly move from the water matrix into a polymer optical sensor located within the cartridge, enabling automated detection by the instrument.[22] These two technologies are a possible alternative to the MPN method, with both using analysis of growth kinetics to estimate numbers of bacteria. These systems present disadvantages such as: (i) quite long delay to obtain a result (i.e. ~6 h when 2,000 *E. coli* are present in 100 mL); and (ii) high equipment cost.

Some recent studies showed that enzymatic based methods could be used to estimate rapidly *E. coli* contamination by measuring β-D-glucuronidase activity without any cultivation step.[6,23,24,25,26] Good correlations in log-log plots of β-D-glucuronidase activity versus *E. coli* numbers enumerated by culture-based methods were generally found for natural water samples. Based on this work, we developed the Coliplage® method which allows the estimation of *E. coli* contamination in bathing water in less than one hour with a detection limit of 100 bacteria per 100 mL.[6] The method is also inexpensive, and the equipment needed is limited to a spectrofluorimeter. Because of its simplicity, this method is accessible to all technicians and can be easily implemented in a laboratory close to the bathing area for active management.[6, 27]

Since 2006, the Coliplage® method has received great interest for real time management of recreational water. Nevertheless, for some identified bathing water types, discrepancies were observed between the standard and Coliplage® methods.[28,29,12,24,25] Therefore, we evaluated alternative methods, including culture with impedance or enzymatic based detection and molecular (PCR) based methods, and their potential performance in bathing water management strategy. The aims of this chapter are: (i) to demonstrate that rapid enzymatic method such as Coliplage® are, in many cases in France, a good tool for active management of faecal contamination, complementary with the reference method (ISO 9308-3); and (ii) to propose, in cases where there are discrepancies, alternative methods to provide a complete bathing water quality assessment strategy.

2 MATERIALS AND METHODS

2.1 Annual bathing water quality monitoring

2.1.1 Sample collection. Samples used in this study were collected in rivers, the English Channel, the Atlantic Ocean and the Mediterranean Sea. Thus 3,431 recreational water samples were collected since 2006 from sites located on the French coastline and from other surface waters (lakes, rivers). All samples were collected in sterile plastic 0.5 L bottles, stored at 4°C, and analyzed within 2h. This sampling procedure is in accordance with the French health organization recommendations.

2.1.2 MPN miniaturized method (reference method) A standardized miniaturized MPN method (ISO 9308-3) using microplate (AES, FR) was used for the enumeration of *E. coli*. In this method, based on the defined substrate approach, 18 mL of sample was added to 18 mL of distilled water or saline buffer (depending on the reported salinity of the sample) to make the first dilution (1/2). A second dilution (1/20) was performed by adding 2 mL of the first dilution to 18 mL of saline buffer. Then 200 µl aliquots of different dilutions (1/2, 1/20) were added to microplate wells containing the substrate, 4-methylumbelliferyl-β-D-Glucuronide (MUG) such that 32 wells contained the 1/20 dilution and 64 wells the 1/2 dilution. The microplate was incubated for 36 to 72 h at 44 °C. The substrate MUG is hydrolyzed by β-D-glucuronidase, releasing the fluorescent compound 4-methylumbelliferone (MUF) that is detectable under ultraviolet light (e.g. 365 nm). *E. coli* abundance was based on the number of positive wells and a statistical analysis using Poisson's law (Most Probable number (MPN) per 100mL). Only the MPN result was considered for this study (without confidence limits). The detection limit of this method is 15 cultivable *E. coli* per 100 mL. The lower and upper limits for each MPN result were obtained from the statistical table provided by the manufacturer.

2.1.3 Coliplage® method. β-D-glucuronidase (GLUase) activity measurements were performed following the protocol proposed by Lebaron et al.,[6] but slightly modified. If turbidity was more than 30 NTU, a prefiltration step was performed using a Nalgene system. Water samples (100 mL) were first filtered through a 47 mm diameter, 20 µm pore-size nylon membrane (Millipore, USA) and then a 10 µm pore-size polycarbonate membrane (Millipore, USA). After that, water samples were filtered through a 47 mm diameter, 0.2 µm pore-size polycarbonate membrane (Millipore, USA). The membranes were placed in 100 ml sterile Erlenmeyer flasks containing 10 ml of sterile phosphate buffer (pH 6.9). The flasks were incubated in a water bath at 44 °C. Next, 2 ml of MUG stock solution (55 mg of MUG (Sigma, USA) and 20 µl of Triton X-100 (Sigma, USA) in 50 ml of sterile water) were added to each flask (final MUG concentration of 0.6 mmol L^{-1}). Every 5 min for 20 min, a 2 ml aliquot was removed and added to a quartz cell and pH adjusted up to 10 by adding 40 µl of a 2N NaOH solution, (this pH corresponds to the fluorescence maximum for MUF).[10] Fluorescence intensities were measured with a fluorescence spectrophotometer (RF 1501, Shimadzu, JP) with an excitation wavelength of 362 nm and an emission wavelength of 445 nm. The spectrofluorimeter was calibrated using a set of MUF (SIGMA, USA) standard solutions from 0 to 500 nM. The production rate of MUF (picomoles of MUF liberated per minute for 100 mL of filtered sample), defined as the GLUase activity, was determined by least-squares linear regression when plotting MUF concentration versus incubation time. The GLUase activities were converted into *E. coli* concentrations by using a straight-line regression from a log-log plot of *E. coli* concentrations estimated by the MPN method versus GLUase activity. Only the estimated value was considered for this study (without confidence rate). A specific straight-line regression was established for each bathing site. An example for marine water samples is presented by Lebaron et al.[6]

2.1.4 The concordance rate evaluation. Currently in France, the Mandatory Value (MV) for managing pollution is 2,000 *E. coli* per 100 mL, but ANSES (Agence Nationale de SEcurité Sanitaire de l'alimentation, de l'environnement et du travail) recommended two thresholds for better consistency between active management of faecal pollution and occurrence of illness: marine water= 1,000 *E. coli* /100 mL and inland water = 1800 *E. coli* /100 mL.[30] Coliplage® is a quality assessment tool, considered as a semi-quantitative method which allows an early evaluation of *E. coli* pollution in water. Coliplage® results must be interpreted as a signal compared to a defined threshold. Thus, results were expressed as an alert signal (red or green) compared to the MV and the ANSES thresholds. Conclusions can be assigned to three cases:

- Concordance case : Coliplage® signal and reference method result were the same compared to the defined threshold, pollution cases (both > MV or ANSES thresholds) and non pollution cases (both < MV or ANSES thresholds))
- Overestimation case : Coliplage® signal > MV or ANSES thresholds > reference method result
- Underestimation cases : Coliplage® signal < MV or ANSES thresholds < reference method result

2.2 Evaluation of alternative methods

2.2.1 Sample collection. All samples (marine water) were collected in sterile plastic 0.5 L bottles, stored at 4°C, and analyzed within 2h. The sampling procedure is in accordance with the French health organization recommendations.

2.2.2 Impedance-based method using BioRad system. The protocol used was provided by the manufacturer BioRad (FR). Each sample (100 mL) was filtered through a 47 mm diameter, 0.45 μm pore-size membrane (Millipore, USA). The membrane was washed with 50 mL of sterile distilled water, then placed in the *E. coli* specific growth media cell (CheckN'Safe™). The cell containing the membrane was placed into the XplOrer64™ apparatus (44°C incubator and reader). The sample was incubated over 9 h (including 1 h preheating). The readings were automatic every 10 minutes after the 1 h preheating. The results were estimated directly by the software (for marine water), and reported as *E. coli* per 100 mL. The analyses were performed on 77 marine water samples collected in 2010.

2.2.3 Enzymatic-based method using ENDETEC system. The protocol used for *E. coli* was provided by the manufacturer ENDETEC (CA) and slightly modified. Each sample (100 mL) was filtered through a 47 mm diameter, 0.45 μm pore-size membrane (Millipore, USA). The membranes were washed with 50 mL of sterile distilled water and placed in the specific *E. coli* growth media cartridge. The cell containing the membrane was placed into the apparatus (41.5°C incubator and reader). The sample was incubated for up to 18 h (including 20 min of preheating). The ENDETEC system continuously monitors a fluorescence signal from the incubating cartridge and reports bacteria as "present" in the sample when the signal crosses a pre-defined threshold. The time when the sample becomes positive is the Time To Detect (TTD). If a sample is not positive when 18 hr is reached, the sample is determined to be "absent" (< 1 VC per 100 mL). The equivalent bacteria concentration (per 100 mL) was automatically calculated using the TTD value and a specific marine water calibration (provided by ENDETEC research department). The analyses were performed on 97 marine water samples collected in 2009 and 2010.

2.2.4 Molecular-based method (qPCR). The protocol used was provided by the manufacturer GeneSystems (FR). Analyses were performed on the GeneSystems apparatus GenExtract and GeneCycler (Recreational GeneDisc). Each sample (250 mL) was set for 30 min at room temperature or 20 min at 70°C. The sample was filtered through a 47 mm diameter, 0.4 µm pore-size polycarbonate membrane. Bacteria were lysed by adding the manufacturer's lysis buffer and DNA extraction and purification was performed by using the GenExtract. DNA was added to the Recreational GeneDisc, and amplification was performed with the GeneCycler. Data was analysed automatically by the software using a specific GeneSystems algorithm, and reported as the Genomic Unit (GU) per volume analysed. As proposed by GeneSystems, UG can be converted for bacteria estimation (per volume analysed) by the relation: 1 GU = 1 bacterium. The analyses were performed on 48 marine water samples collected in 2009.

3 RESULTS

3.1 Comparison between the reference method and the Coliplage® method: a follow up for 5 years

Since 2006, more than 70 beaches have been monitored each summer. The bathing areas were located in the English Channel, the Atlantic Ocean, the Mediterranean Sea and various lakes and rivers in France. Different site classifications could be identified such as sandy beaches, proximity to a marina, proximity to a waste water discharge, etc. More than 12,000 Coliplage® results were collected in France including 3,428 samples analyzed with both Coliplage® and the reference method (MPN).

The results presented in Table 1 compared the two methods according to the 1976 council Directive MV and the two ANSES thresholds. Since 2006, the number of samples analysed with the two methods has been increased.

For the MV of 2,000 bacteria per 100 mL (A), the concordance rate remained stable each year, with an average concordance percentage around 92.95% for 5 years. The percentage of pollution cases detected by the both methods had a tendency to decrease since 2006 from 8.11% to 2.80%. Correspondingly, the percentage of non pollution cases slightly increased since 2006. Concerning the non-concordance cases, we can classify them into 2 categories: Coliplage® > MV > MPN, which represents overestimation of *E. coli* contamination by Coliplage® and Coliplage® < MV < MPN, which is underestimation. For the first category, the percentage of overestimation has slightly increased since 2006. Conversely, the second category has had a tendency to decrease.

Concerning the ANSES thresholds of 1,000 or 1800 bacteria per 100 mL (B), the concordance rate is close to 88% for the 5 years, and we observed a slight decrease of around 5% compared with the percentage obtained with MV at 2,000 bacteria per 100 mL. The rates of overestimation of *E. coli* levels by Coliplage® compared to reference method is still higher than underestimation whatever the threshold considered.

Table 1 *Concordance rates between the Coliplage® and reference (MPN) methods compared table **A)** the Council Directive MV = 2,000 bacteria per 100 mL and **B)** the ANSES thresholds = 1,000 or 1800 bacteria per 100 mL. n= 3,428.*

A

Year	Numbers of analyses per year	Concordance		Non concordance	
		Pollution cases	Non pollution cases	Coliplage > MV > MPN	Coliplage < MV < MPN
2006	481	8,11%	83,99%	4,57%	3,33%
2007	380	6,58%	85,00%	5,26%	3,16%
2008	849	4,95%	87,99%	5,65%	1,41%
2009	749	2,80%	89,32%	6,28%	1,60%
2010	972	2,88%	91,67%	4,01%	1,44%
	Weighted mean	**4,52%**	**88,43%**	**5,13%**	**1,92%**
	Confidence	*+/- 2,7%*	*+/- 3,7%*	*+/- 1,2%*	*+/- 1,1%*
	TOTAL	**92,95%**		**7,05%**	

B

Year	Numbers of analyses per year	Concordance		Non concordance	
		Pollution cases	Non pollution cases	Coliplage > threshold > MPN	Coliplage < threshold < MPN
2006	478	10,67%	74,27%	8,16%	6,90%
2007	380	11,58%	75,00%	6,05%	7,37%
2008	849	8,36%	81,27%	8,24%	2,12%
2009	749	4,14%	84,51%	8,81%	2,54%
2010	972	3,70%	84,57%	8,64%	3,09%
	Weighted mean	**6,80%**	**81,24%**	**8,23%**	**3,73%**
	Confidence interval	*+/- 4,3*	*+/- 5,7*	*+/- 1,1*	*+/- 2,8*
TOTAL		**88,04%**		**11,96%**	

3.2 Summary of alternative methods

The comparison of advantages and disadvantages of each selected method is presented in Table 2.

Some criteria were more important than others to classify methods of interest for active management, such as specificity (viability) and time to get a result. The position of each method regarding these two criteria is illustrated in Figure 1.

Table 2 *E. coli detection methods tested in this study.*

	Culture based method	Enzymatic method (no culture)	PCR method (no culture)	Automatic culture and metabolic based method	
	ISO 9308-3	Coliplage®	Genesystems	ENDETEC	Biorad
Steps	- Sample dilution - Microplate loading - Incubation - Manual reading	- Filtration - 30 min for reading	- Filtration - Extraction & purification - Disc loading - Automatic reading	- Filtration - Incubation - Automatic reading	- Filtration - Incubation - Automatic reading
Delay (Analysis time)	36-72H	1 hour	4 hours	4-18 hours depending on bacteria concentration	4-10 hours depending on bacteria concentration
Specificity concerning					
Dead bacteria	Not Detected	Not Detected	Detected	Not Detected	Not Detected
ANC bacteria	Not Detected	Detected	Detected	Not Detected	Not Detected
Non specific bacteria	Not Detected	Few marine *Vibrio*	Not Detected	Not Detected	Not Detected
Sensibility	15 bacteria /100mL	100 bacteria /100mL	75 GU /100mL	1 bacterium /100mL	41 bacteria /100mL
Implementation	local lab	local lab	central lab	local lab	local lab
operator background	non experienced staff	non experienced staff	experienced staff	non experienced staff	non experienced staff
Cost					
appartus	3-5k€	10-15k€	30-40k€	11-15k€	50-70k€
consumables	10-15€	10-15€	10-20€	7-11€	10-15€
Number of samples per analysis	1	6	4	16	64
Limitations of use	Not identified	Presence of UV desinfected wastewater	Presence of UV desinfected wastewater turdib water	Not identified	Not identified

As mentioned in Table 2 and illustrated in Figure 1, concerning the time to result, the methods can be classified from most to least rapid as follows: Coliplage® > molecular based method (PCR) > automatic culture and metabolic based methods > culture based method (reference method).

Considering the viability of bacteria, culture and metabolic based methods appear to be the more adapted to detect only viable and cultivable bacteria, in agreement with the reference method. However, as reported in Table 2, the time to get a result is longer (4 to 10-18h) than specified in the 2006 Council Directive requirements (for active management). To summarize, the more the method is rapid the less it tends to be specific.

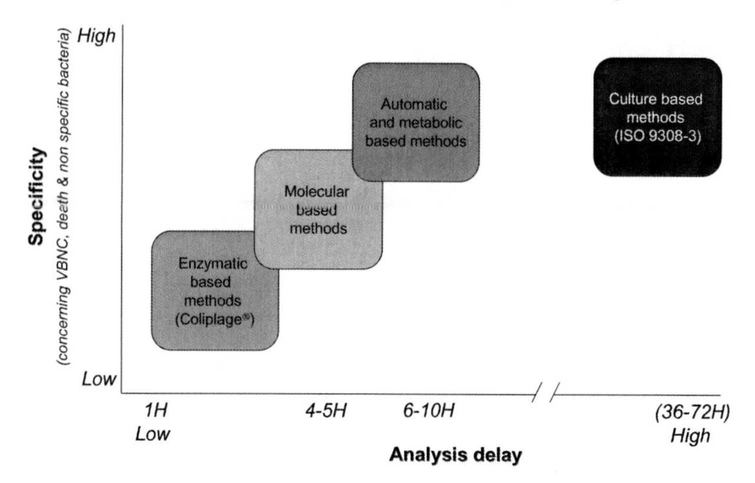

Figure 1 *Schematic representation of efficiency of each method for bathing water management.*

3.3 *In situ* evaluation of alternative methods

The results of the three evaluation campaigns of *E. coli* alternative methods in bathing sites (marine water) are presented in Table 3.

Table 3 *Comparison of concordance rates (%) for different MV (**A**= 2,000 and **B** = 1,000 E. coli/ 100 mL), between the reference method (ISO 9308-3) and alternative methods during 3 different campaigns: Coliplage®/GeneSystems; Coliplage®/Impedance and Coliplage®/indicator enzyme (ENDETEC system).*

A		Conclusions cases		
		Concordance	Overestimation	Underestimation
n=48	Coliplage®	68,75%	25,00%	6,25%
	PCR	62,50%	31,25%	6,25%
n=77	Coliplage®	81,82%	12,99%	5,19%
	Impedance	94,81%	2,60%	2,60%
n=97	Coliplage®	72,00%	23,00%	5,00%
	indicator enzyme	89,00%	6,00%	5,00%

B		Conclusions cases		
		Concordance	Overestimation	Underestimation
n = 48	Coliplage®	50,00%	43,75%	6,25%
	PCR	60,42%	29,17%	10,42%
n=77	Coliplage®	80,52%	14,29%	5,19%
	Impedance	97,40%	1,30%	1,30%
n=97	Coliplage®	58,00%	34,00%	8,00%
	indicator enzyme	90,00%	6,00%	4,00%

Whatever the analysis campaign or the mandatory value, we observed in Table 3 that Coliplage® presents some discrepancies (respectively 18.18% to 31.25% and 19.48% to 50% for the MV at 2,000 and the ANSES threshold at 1,000 *E. coli* / 100mL) compared to the reference method, especially over-estimation cases. This confirms the previous conclusions that application of Coliplage® is limited in certain bathing sites, and thus we can conclude that the samples used in this study are relevant for evaluating the efficiency of alternative methods in comparison to Coliplage®. Note that samples used in the impedance (BioRad) campaign have presented fewer discrepancies compared to samples used for the two other campaigns (qPCR and indicator enzyme (ENDETEC system)), so conclusions concerning the efficiency of this method could be affected.

For the MV of 2,000 *E. coli* per 100 mL (3A), the indicator enzyme based on the ENDETEC technology and impedance method, which were expected to have greater specificity, gave a higher concordance rate with the reference method compared to Coliplage®: 89% (+17%) for the indicator enzyme and up to 94% (+13%) for the impedance method. The PCR method was less effective, with a decrease of concordance to 62.5% (-6.3%) compare to the two previous methods and Coliplage®. In conclusion, the indicator enzyme and impedance methods seem to be the more accurate methods for this threshold.

For the ANSES threshold at 1,000 bacteria per 100 mL (3B), the indicator enzyme and impedance methods also gave a higher concordance rate with the reference method compared to Coliplage®: 90% (+32%) for the indicator enzyme and up to 97% (+17%) for the impedance method. The PCR method was less effective compared to the two previous methods, with a concordance of 60.4% (+10.4%). The most accurate method (for the ANSES threshold at 1,000 bacteria per 100 mL) seems to be the indicator enzyme method,

As presented in Tables 1 and 3, whatever the threshold (MV or ANSES) and method, concordance with the reference method was not perfect, and differences between results will always be present (over or under estimation). These results show that each method presents limitation, especially concerning specificity (VBNC, dead or non specific bacteria).

4 DISCUSSION

In this article, we presented a Coliplage® data compilation collected over five years from more than 100 different bathing sites. We compared the reference MPN method used to detect *E. coli* in France with the Coliplage® method to assess the use of this rapid enzymatic method in the daily active management of recreational water.

Coliplage® has confirmed over 5 years that this enzymatic based method has a great potential to monitor faecal pollution in bathing water due to its rapidity (1 hour), simplicity and cost compared to the conventional method or other alternative methods. In more than 92% of measurements (Table 1), this method provided great concordance with the reference method, and even when the new ANSES threshold is considered, the percentage is close to 87%. Thus the Coliplage® method offers a useful tool for an active faecal pollution management by local authorities and to rapidly monitor the microbiological bathing water quality, as a complement to the reference method. However, we observed a slight number of samples (near 7%) which presented different conclusions between Coliplage® and the reference method. These discrepancies were described previously in the literature.[12] We can underline three major reasons to explain these slight differences.

For the cases of underestimation by Coliplage®, there is the potential presence of *E. coli* which does not express β-D-glucuronidase in natural water,[31,32,25,33] whereas in culture conditions like the MPN method these bacteria will have enzymatic activity induced by the medium.. We performed assays with high concentrations of *E. coli* from pure culture in BHI medium (10^6 bacteria per mL) in Coliplage® and the reference method (data not shown), and no GLUase activity was measured by Coliplage®, while quantification with the reference method was observed. More experiments need to be performed to understand if this is a caused by non-detection of the glucuronidase activity, or reduced induction of the enzyme in the target cells.

For the cases of over-estimation, the non-concordance can be attributed to the presence of VBNC cells and non specific cells. The VBNC state is defined as cells that will not grow on a favourable culture medium but keep some metabolic activity. This kind of state can be induced by stress from environmental parameters such temperature, salinity, UV, O_2, predation, competition with other bacteria, and lack of nutriments in the environment.[34,35,36] These forms of stress are found in the natural environment, and in order to estimate more accurately the sanitary risk associated with bathing water, it is important to considerer the presence of VBNC.[37] Indeed, the presence of VBNC cells attests to the presence of faecal pollution with a risk of associated pathogens (such as parasites and viruses) that is impossible to detect with culture-based methods including the reference method.[7,38] Coliplage® appears to be the most rapid and pertinent method which can detect VBNC. The over-estimation of GLUase activity was also previously described by many authors, suggesting that other members of *Enterobacteriaceae* including some *Shigella* and *Salmonella* stains, a few *Yersinia, Citrobacter, Edwardia* and *Hafnia* strains, and other non *Enterobacteriaceae* strains (*e.g.* a few marine *Vibrio* species) could potentially contribute to this enzymatic activity.[39,13,12]

To remedy this lack of robustness, we tried to identify one or more alternative methods to allow us to maintain a daily follow-up of the bathing water quality on all the French bathing sites. We identified three complementary methods: the molecular (PCR) method, the impedance method and the indicator enzyme method. As presented in this paper, it seems that if we want to have a higher correlation with the reference method, we need to use culture and metabolic methods (impedance or indicator enzyme), with a recommendation for the ENDETEC indicator enzyme method which had better efficiency compared to the impedance-based method, especially for the ANSES threshold. In fact, interfering bacteria are inhibited by the culture step, which increases selectivity. As a consequence, however, time-to-result is longer and may not meet the 2006 council Directive expectations. PCR could be an alternative, but difficulties in implementation and specificity of results seem to be the two parameters limiting its use for bathing water quality management.

5 CONCLUSIONS

It appears that with the 2006 Council Directive that reference method (ISO 9308-3) is not applicable for active (real–time) faecal pollution management due to its time-to-result (2 days). Alternative methods currently available could be good complementary tools for active faecal pollution management. Coliplage® due to its rapidity (1 hour), low cost and easy implementation, is the most accurate method to determine the microbiological water quality, and to allow an active management of the bathing area as a complement to reference method. However, Coliplage® and the reference method show some differences in less than 10% of cases. Concerning Coliplage®, discrepancies appeared on particular bathing sites such as close to a port, a UV disinfection water treatment plant discharge, or a marine pond (tidal water change). Concerning other alternative methods: i) PCR is a rapid method (4 hours) but it overestimates *E. coli* contamination especially in sites which present recurrent dead and VBNC *E. coli* cells which contain DNA, ii) Automatic culture and metabolic based methods seem to be the more efficient compared to ISO 9308-3, with a higher concordance correlation (to 90%) compared to Coliplage®, especially in sites with interfering bacterial population (*i.e.* VBNC, dead or non specific). For the majority of bathing sites, Coliplage® is the most efficient method. For the sites with problems in terms of VBNC, dead cells and non specific bacteria, automatic and metabolic based methods could be recommended for the specific detection of *E. coli*.

Acknowledgements

This research was supported by Veolia Environnement Research and Development. We wish to thank scientists from ENDETEC (especially Eric MARCOTTE and Bradley TAYLOR), in Kingston (Canada), for their contributions and help for this research. We wish to thank VEOLIA Eau laboratories (France) and the Veolia GIRAC project manager (Bretagne) for their contributions of collecting the samples and performing the filtration and analyses.

References

1. EP/CEU (European Parliament/Council of the European Union), 1976,
2. The European Union, *official journal of European Union*, 2006,
3. S. C. Edberg, E. W. Rice, R. J. Karlin, and M. J. Allen, *Symp. Ser. Soc. Appl. Microbiol.*, 2000, **29**, 106S.
4. New York State Department of Health, 2011,
5. T. Garcia-Armisen, P. Lebaron, and P. Servais, *Lett Appl Microbiol*, 2005, **40**, 278.
6. P. Lebaron, A. Henry, A. S. Lepeuple, G. Pena, and P. Servais, *Mar.Pollut.Bull.*, 2005, **50**, 652.
7. G. Caruso, E. Crisafi, and M. Mancuso, *J Appl Microbiol*, 2002, **93**, 548.
8. M. Kilian and P. Bulow, *Acta Pathol Microbiol Scand.*[B], 1979, **87**, 271.
9. ISO, *Microbiological methods*, 1998,
10. I. George, M. Petit, and P. Servais, *J.Appl.Microbiol.*, 2000, **88**, 404.
11. C. J. Palmer, Y. L. Tsai, A. L. Lang, and L. R. Sangermano, *Appl Environ Microbiol*, 1993, **59**, 786.
12. J. Baudart, P. Servais, P. H. De, A. Henry, and P. Lebaron, *J Appl Microbiol*, 2009, **107**, 2054.
13. J. M. Pisciotta, D. F. Rath, P. A. Stanek, D. M. Flanery, and V. J. Harwood, *Appl Environ Microbiol*, 2002, **68**, 539.
14. R. T. Noble and S. B. Weisberg, *J.Water Health*, 2005, **3,** 381.
15. A. Rompre, P. Servais, J. baudart, M. R. de-Roubin, and P. Laurent, *J Microbiol Methods*, 2002, **49**, 31.
16. M. Ruan, C. G. Niu, P. Z. Qin, G. M. Zeng, Z. H. Yang, H. He, and J. Huang, *Anal.Chim.Acta*, 2010, **664**, 95.
17. J. Baudart and P. Lebaron, *J.Appl.Microbiol.*, 2010, **109,** 1253.
18. J. Dupont, F. Dumont, C. Menanteau, and M. Pommepuy, *J Appl.Microbiol*, 2004, **96**, 894.
19. P. Galant, *Faecal indicators: problem or solution? Proceedings*, 2011,
20. S. Timms, K. O. Colquhoun, and C. R. Fricker, *J Microbiol Methods*, 1996, **26**, 125.
21. Bio-Rad, *Xplorer catalog*, 2009,

22. ENDETEC *website*, 2011,
23. D. Wildeboer, L. Amirat, R. G. Price, and R. A. Abuknesha, *Water Research*, In Press, Corrected Proof,
24. P. Bergeron and S. Courtois, *L'eau, L'industrie, les nuisances*, 2009, **313**, 92.
25. L. Fiksdal and I. Tryland, *Curr.Opin.Biotechnol.*, 2008, **19**, 289.
26. P. Servais, T. Garcia-Armisen, A. S. Lepeuple, and P. Lebaron, *Annals of Microbiology*, 2005, **55**, 151.
27. L. Fiksdal, M. Pommepuy, M. P. Caprais, and I. Midttun, *Appl Environ Microbiol*, 1994, **60**, 1581.
28. S. Abboo and B. I. Pletschke, *Water SA*, 2010, **36**, 133.
29. M. Asma, A. Badshah, S. Ali, M. Sohail, M. Fettouhi, S. Ahmad, and A. Malik, *Transition Metal Chemistry*, 2006, **31**, 556.
30. ANSES (AFSSET), 2007,
31. G. W. Chang, J. Brill, and R. Lum, *Appl Environ Microbiol*, 1989, **55**, 335.
32. P. Feng, R. Lum, and G. W. Chang, *Appl.Environ.Microbiol.*, 1991, **57**, 320.
33. A. H. Farnleitner, L. Hocke, C. Beiwl, G. C. Kavka, T. Zechmeister, A. K. Kirschner, and R. L. Mach, *Lett Appl Microbiol*, 2001, **33**, 246.
34. A. Villarino, A. L. Toribio, B. M. Brena, P. A. Grimont, and O. M. Bouvet, *Biotechnol.Lett.*, 2003, **25**, 1329.
35. A. Villarino, M. N. Rager, P. A. Grimont, and O. M. Bouvet, *Eur.J.Biochem.*, 2003, **270**, 2689.
36. M. Gourmelon, Safe Management of Shellfish and Harvest *Waters.*, 2010,
37. S. M. Ben, O. Masahiro, and A. Hassen, *Ann.Microbiol.*, 2010, **60**, 121.
38. A. Rompre, P. Servais, J. Baudart, M. R. de-Roubin, and P. Laurent, *J Microbiol Methods*, 2002, **49**, 31.
39. C. M. Davies, S. C. Apte, S. M. Peterson, and J. L. Stauber, *Appl. Environ. Microbiol.*, 1994, **60**, 3959.

DO BIOFILMS DEVELOPED IN THE RIVER BED SERVE AS SOURCES FOR BACTERIAL INDICATORS?

H. Hirotani and M. Yoshino

Division of Natural Sciences, Osaka Kyoiku University, Kashiwara, Osaka, 582-8582, Japan

1 INTRODUCTION

Biofilms are developed in riverbeds form complex communities consisting of diverse microorganisms. The characteristics of biofilms are affected by past and current environmental conditions and, they influence the surrounding river water quality. Formation of river biofilms and their activities have been studied, including their nutrient interactions[-3], response to flow velocity, [4,5] and their interaction with microbial indicators. [6,7]

To evaluate the environmental condition of the river, it is a common practice to sample the flowing water for analyses. In Japan, total coliforms serve as a water quality standard indicating faecal pollution for river, lake and ocean water. Since total coliform counts inevitably contain false positive results, more than 70% of the locations still do not meet the coliform standard (http://www.env.go.jp/water/suiiki/h15/index.html). Today, it is widely known that total coliforms include bacteria not derived from faecal contamination and that wild animals contribute to the elevated faecal indicator concentrations in regulated surface and ground waters.[8] *Escherichia coli* is another indicator of faecal contamination of water, however *E. coli* are known to reproduce in the environment in the tropics. [9,10]

Since significant populations of microorganisms inhabit river biofilms, monitoring for microorganisms only in the flowing water may misinform the river environment examination results. Biofilms present on the surface of pebbles in the riverbed may serve in part as a source of replicating indicator microorganisms in the river water, which possibly can result in false-positive indicator values. On the other hand, the detection of indicator bacteria in biofilms may merely indicate the presence of contamination in the riverbed, supposing that the indicators are of allochthonous origin. In the present study, field sampling was undertaken six times to measure the microbial indicators present in natural riverbed biofilms as well as in flowing water. The aim of this study was to investigate the relation of indicator bacteria in the water column and in the river bed biofilm, and to determine if the biofilms can serve as the source of the indicator bacteria in river water.

2 MATERIALS AND METHODS

2.1 Study Site and Sampling

The study was undertaken in the Yamatogawa watershed, in Osaka, Japan. Water samples and pebbles were collected on 2 Feb., 4 Apr, 18 Jun., 6 Aug., 20 Sep., and 25 Oct., 2007, from five sampling stations. The sampling stations were located from mountain streamsin a wooded area of quasi-national park and from to an urbanised reiver corridor receiving municipal wastes (Figure 1). Water samples were taken directly into sterile polypropylene bottles. Pebbles, which were totally exposed and not submerged in the sediments, were grab-sampled from the river bed and stored in Ziploc bags (SC Johnson & Son, Racine, WI, USA). The samples were brought to the laboratory within 3 h on ice.

Biofilms developed on the surface of the pebbles were scrubbed off with nylon toothbrushes into sterile water in the laboratory. The biofilm samples and the water samples were subjected to the microbial analyses. The surface areas of the pebbles were estimated by the following formula: [5]

$$\text{Surface area (cm}^2\text{)} = 3.51 \text{ weight (g)}^{2/3}$$

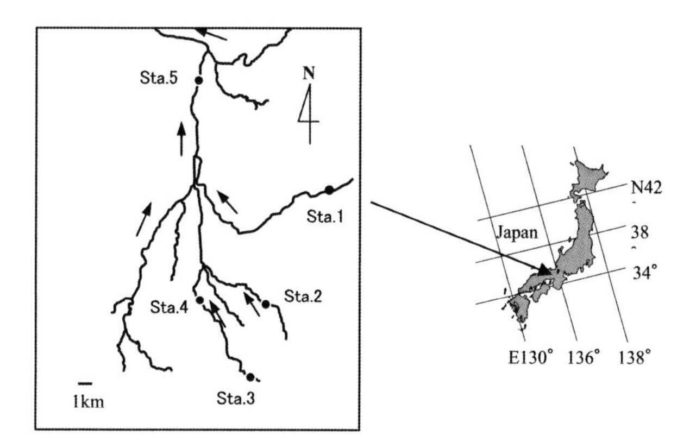

Figure 1 *The Yamatogawa watershed and the sampling stations.*

2.2 Sample Analyses

Heterotrophic bacteria were enumerated using R2A plates, incubated for 5 d at 20°C. [11] Total coliforms and *E. coli* were enumerated by the MPN method using Colilert 18 and Quanti-Tray/2000 (Idexx, Westbrook ME, USA). Total coliforms and *E. coli* in water are commonly expressed as numbers per 100 mL, but in this study it is expressed per mL for easier comparison of observed biofilm surface densities (per one cm^2).

3 RESULTS AND DISCUSSION

3.1 Heterotrophic Bacteria

Heterotrophic bacterial densities ranged from 2.4×10 to 2.2×10^5 cfu/ml in river water and 9.7×10^2 to 3.4×10^6 cfu/cm^2 in biofilms. Heterotrophic bacteria generally increased downstream both in water and in biofilms. The thicknesses of the biofilms developed on the sampled pebbles were not uniform between samples, but it was considered that the observed community compositions were mature since the pebbles were in place more than few days.[12]

Heterotrophic counts in biofilms significantly correlated ($r = 0.65$, $n = 30$, $P < 0.01$) with bacteria in water columns sampled at the same location and time (Figure 2). Almost all the plots in the figure distributed above the broken line, which indicates the identical value. Plots falling on the broken line mean that biofilm contained the same amount of bacteria within the 1 cm thickness of the water above the surface of the pebble ($cm^2 \times cm = mL$). Not surprisingly, biofilms generally contained bacteria in greater densities than the river water.

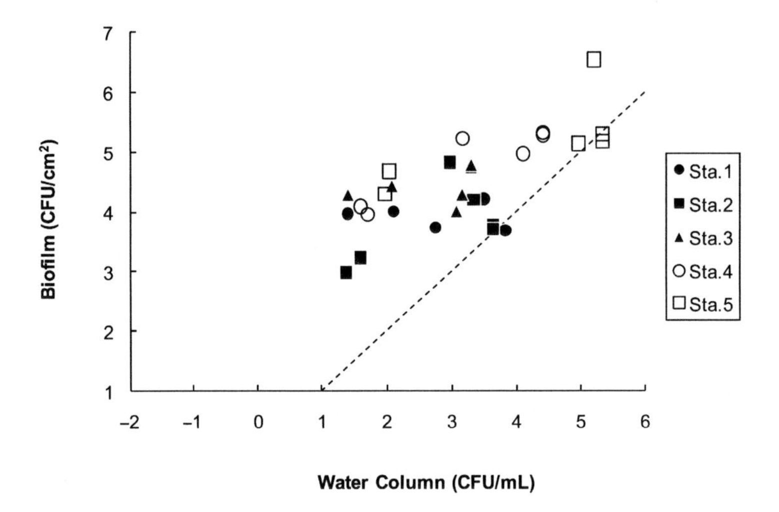

Figure 2 *Heterotrophic counts in biofilm and in water column.*

Correlation of heterotrophic bacteria in biofilm with river discharge has been reported.[7] This relation suggests the growth regulation of the bacteria in biofilm by nutrient supply from flowing water. If nutrients from the river water promote the growth of heterotrophic bacteria in biofilm, it is suggested that bacteria in biofilm serve as the source of bacteria in the water column.

3.2 Total Coliforms

Total coliforms are the most commonly used indicator for faecal contamination of water, but their validity has been questioned elsewhere.[8] Densities of total coliforms below suburban areas always exceeded those at pristine stations. The correlation of total coliforms in water and biofilm (Figure 3) was significant ($r = 0.60$, $n = 30$, $P < 0.01$)

3.3 *Escherichia coli*

E. coli in water is considered to be a more appropriate faecal indicator than total coliforms in indicating the presence of faecal contamination in river water.[13] Some of the data points spread below the broken line in Figure 4 indicating a greater proportion of *E. coli* in water than biofilm for these observations. Relatively greater concentrations of *E. coli* were observed at stations 4 and 5, which received heavier contamination loads than at upstream rural sampling stations. Although the overall correlation of *E. coli* in the water column and in the biofilm was significant ($r = 0.50$, $n = 30$, $P < 0.01$), the scatterplot seems not be indicative of the relation between the bacteria in two habitats.

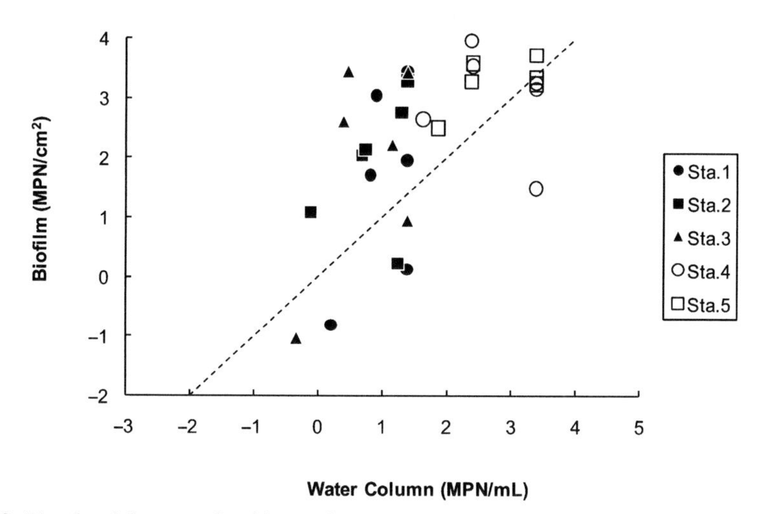

Figure 3 *Total coliforms in biofilm and in water column.*

If a significant amount of bacteria in the biofilms was derived from the bacteria in the water column, or *vice versa*, the increased densities in the water in the suburban stations should result in increased densities in the biofilm at the same locations. Since the increase of *E. coli* in biofilms did not follow the observed increase in the water columns, the attachment of planktonic *E. coli* to biofilm was considered negligible in the river environment.

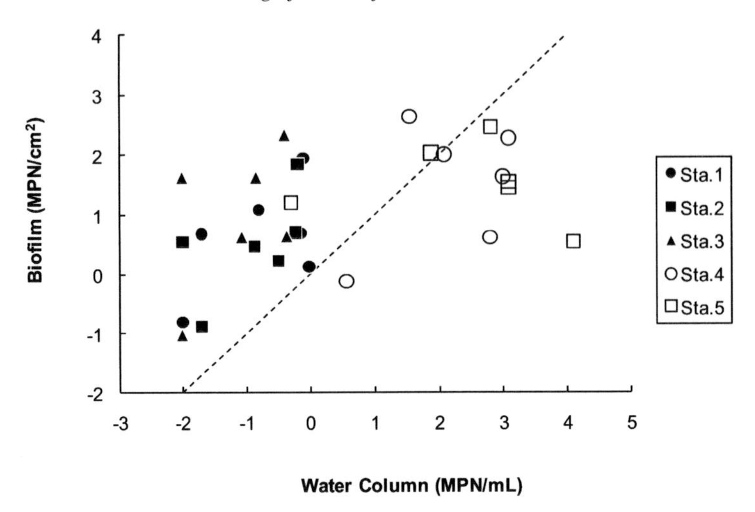

Figure 4 Escherichia coli *in biofilm and in water column.*

A positive and significant correlation has been found between water temperature and *E. coli* in the biofilm. [7] The positive thermal effect of water temperature suggests the growth of *E. coli* in the biofilm. Detection of *E. coli* in biofilms may not directly mean the presence of faecal contamination in the river. Stream sediments have been recognized as in-channel store of faecal contamination.[14] If *E. coli* can multiply in biofilm, and are released to the river water, biofilm can be said to serve as a source of the indicator bacteria in the water column causing a false positive *E. coli* count. However, the external load of faecal contamination may indicate a greater effect on *E. coli* densities below the suburban locations.

Although in marine water *E. coli* is considered to die-off easily,[15] riverbed biofilms may serve as a refuge area for the bacteria which enable longer survival in the river environment. Biofilm may block lethal sunlight from killing the attached bacteria and may protect the bacterial flora from occasional winter freeze. In the suburban areas, the external faecal indicator input may be heavy enough to obscure the internal load of indicator which is derived from the in-situ biofilm. However, in the relatively clean upstream wooded areas, significant indicator concentrations, which show positive correlation with water temperature, may derive from biofilms.

4 CONCLUSION

Microbial indicators in naturally developed riverbed biofilms were investigated in pristine wooded streams which then passed through an urban corridor. Indicator bacteria generally indicated exhibited elevated concentrations downstream in streambed biofilm as well as in the water column. Heterotrophic bacteria in biofilm indicated a positive correlation with the river discharge, suggesting an involvement of nutrient availability for the growth of bacteria in biofilm. Detection of *E. coli* in biofilm may not directly mean the presence of faecal contamination in the river water. The increase of *E. coli* downstream was greater in the water column than biofilm. It was considered that the attachment of planktonic bacteria to biofilm was negligible. In the river water quality assessment, one must be aware that riverbed biofilm can serve as a source of indicator bacteria causing a false-positive result especially in the pristine areas.

Acknowledgements

This study was supported in part by the River Fund in charge of the Foundation of River and Watershed Environment Management (FOREM), Japan.

References

1 S.A Ble nkinsopp and M.A. Lock M. A., *Wat. Res.*, 1990, **24**, 441.
2 M.N. Mohamed, J.R. Lawrence and R.D. Robarts, *Microb. Ecol.*, 1998, **36**, 121.
3 L. Hall-Stoodley, J.W. Costerton and P. Stoodley P, *Nature Rev. Microbiol.*,2004, **2**, 95.
4 T.J. Battin, L.A. Kaplan, J.D. Newbold, X. Cheng and C. Hansen, *Appl. Environ. Microbiol.*,2003 **69**, 5443.
5 H. Hirotani, Y. Yamazoe, N. Kunimoto and M. Yoshino, Biofilms III: 3rd International Conference, München, 2008, *Abstracts*, 122.
6 M. Balzer, N. Witt, H.C. Flemming, J. Wingender, *Wat. Sci. Tech.*, 2010, **61**, 1105.
7 H. Hirotani and M. Yoshino, *Wat. Sci. Tech.*, 2010, **62**, 1149.
8 G.A. Toranzos, G.A. McFeters and J.J. Borrego, Manual of Environmental Microbiology, C. J. Hurst, R. L. Crawford, G. R. Knudsen, M. J. McInerney, and L. D. Stetzenbach (eds.), 2nd edn, ASM Press, Washington DC, 2002, 205.
9 M. Carrillo, E. Estrada and T.C. Hazen, *Appl Environ Microbiol*, 1985, **50**, 468.
10 R.S. Fujioka and L.K. Shizumura, *J. Water Pollut. Control.Fed.*, 1985, **57**, 986.
11 APHA, Standard Methods for the Examination of Water and Wastewater, 20th edn, Washington DC, 1998.
12 R. Araya, K. Tani, T. Takagi, N. Yamaguchi and M. Nasu, *Microbial. Ecol.*, 2006, **43**, 111.
13 H. Hirotani and M. Yoshimura, 1st IWA-ASPIRE (Asia Pacific Regional Group) Conference and Exhibition, 2005, Proceedings, 6F-1.
14 R. W. Muirhead, R. J. Davies-Colley, A. M. Donnison and J. W. Nagels, *Wat. Res.*, 2004, **38**, 1215.
15 J. J. Borrego, F. Arrabal, A. de Vicente, L. F. Gomez and P. Romero, *J. Water Pollut. Control. Fed.*, 1983, **55**, 297.

COST-EFFECTIVE APPLICATIONS OF HUMAN AND ANIMAL VIRUSES AS MICROBIAL SOURCE-TRACKING TOOLS IN SURFACE WATERS AND GROWDWATER

Sílvia Bofill-Mas, Byron Calgua, Jesus Rodriguez-Manzano, Ayalkibet Hundesa, Anna Carratala, Marta Rusiñol, Laura Guerrero, Rosina Girones

University of Barcelona, Department of Microbiology, Diagonal, 635, 08028-Barcelona, Spain

1 INTRODUCTION

Viruses are found in virtually all natural aquatic environments, where they reach a concentration of between 10^7-10^8 virus-like particles per mL, which is higher than that of bacteria.[1] Even in highly industrialized countries, viruses that infect humans prevail throughout the environment, causing public health concerns and leading to substantial economic losses. Our studies of the DNA viruses adenoviruses (AdV) and polyomaviruses (PyV), excreted by humans and widely prevalent in urban wastewater, has changed, in our view, the paradigm of viral contamination previously considered as a sporadic event related to the presence of outbreaks in the population. The accumulated data on adenoviral and polyomaviral infections support the concept of viruses as members of the human microbiome and their use as molecular markers of faecal/urine contamination in water, food and the environment. In fact, many human viruses could be considered part of the human microbioma and some of them have shown to offer protection against infections by other pathogens.[2]

Many orally transmitted viruses produce subclinical infection and symptoms in only a small proportion of the population. However, some viruses may give rise to life-threatening conditions, such as acute hepatitis in adults, as well as severe gastroenteritis in small children and the elderly. The development of disease is related to the infective dose of the viral agent, the age, health, immunological, and nutritional status of the infected individual (pregnancy, presence of other infections or diseases), and the availability of health care. Potentially pathogenic viruses in urban wastewater include HAdV and HPyV, which are detected in all geographical areas and throughout the year, and enteroviruses (EV), NoV, RV, astroviruses, HAV and HEV, with variable prevalence in different geographical areas and/or periods of the year.

Studies in our laboratory suggest that HAdV has considerable potential as an indicator of human faecal contamination and that JCPyV is a complementary human-specific parameter. Specific animal PyV and AdV have also been identified and we have proposed the use of specific HAdV[3,4] and HPyV[5,6], porcine adenoviruses (PAdV)[7,8,9] and bovine polyomaviruses (BPyV)[8,9,10] as new indicators of faecal/urine contamination and as microbial source tracking (MST) tools.

Those viruses that are transmitted *via* contaminated food or water are typically stable because they lack the lipid envelopes that render other viruses vulnerable to environmental agents. Some viruses, such as human polyomaviruses and adenoviruses (HAdV), infect humans during childhood, thereby establishing persistent infections. In the case of some AdV that infect the respiratory tract, many viral particles may be excreted in feces for months or even years.[11] Furthermore, HAdV 1, 2, 5, 7, 12 and 31, among others, have been detected in contaminated water and shellfish[12,13] in addition to HAdV 40 and 41, which produce gastroenteritis and are excreted by children with a very high concentration in feces (10^{11} viral particles per gram) .

HAdV are found in nearly 100% of urban wastewater samples tested, including those from cities in Africa, the United States of America, Central and South America and Europe. Adenoviruses are also frequently detected in shellfish, including samples that met current safety standards, based on levels of faecal bacteria.[13] Officials at the U.S. Environmental Protection Agency recently added AdV to the list of potentially dangerous contaminants, a move that was based on their high prevalence in urban wastewater, their ability to cause disease, and their resistance to water purification treatments, particularly those involving UV irradiation.

Efficient qPCR assays have been developed for the detection and quantification of HAdV[12,14] and JCPyV.[15] Nested-PCR and nucleotide sequence analyses of wastewater samples have revealed the large diversity of viruses that are responsible for clinical and subclinical infections in urban populations. These analyses also help investigators to characterize such viruses genetically, even those that can not be cultured efficiently in cell lines, including the JC polyomavirus (JCPyV), hepatitis E virus (HEV) and noroviruses (NoV).[5,16,13] Genomic amplification techniques also generate more comprehensive information on hepatitis A virus (HAV) strains, rotaviruses (RV), and those AdV that remain difficult to isolate in cell cultures.

Advances in concentration methods for viruses in water and molecular assays, such as quantitative PCR (qPCR), provide sensitive, rapid, and quantitative analytical tools with which to study viruses in water and to develop new standards for improving the control of the microbiological quality of food and water, to trace the origin of faecal contamination, and to assess the efficiency of virus removal in wastewater treatment plants.

2 METHODS USED TO CONCENTRATE VIRUSES IN WATER

A wide range of concentration methods have been described to recover viruses from urban and slaughterhouse wastewater, well water, lakes, the sea and river water. These methods seek to concentrate viruses from large volumes to smaller volumes ranging from 10 mL to 100 µL. Most of the methods used for river and sea water are based on adsorption-elution processes using membranes, filters or matrixes like glass wool.[17,18,19] However, they are two-step methods that can be cumbersome and could hamper the simultaneous processing of a large number of samples. Moreover, in these methods, clogging may occur in samples with high concentrations of organic matter. In addition, low viral recovery values have been reported for sea water samples, as shown in Figure 1.[19]

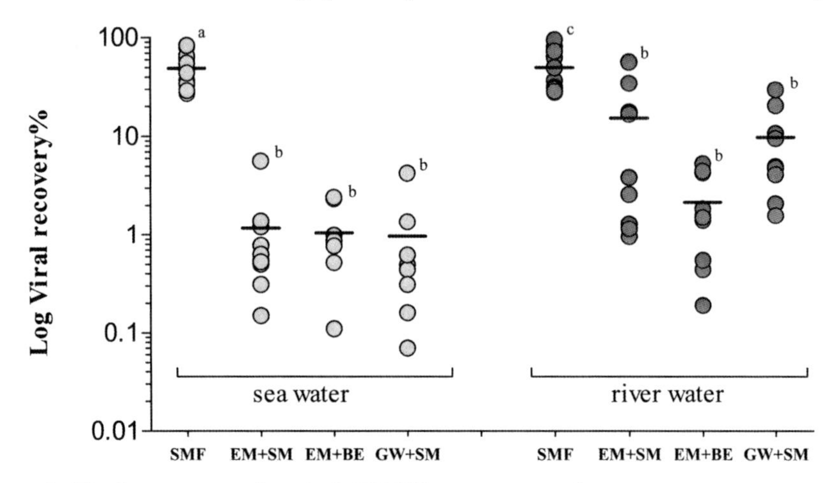

Figure 1 *Viral recovery of spiked HAdV2 in river and sea water concentrated by different methods, and detected and quantified by TaqMan® qPCR. SMF: One-step skimmed milk flocculation-based method; EM+SM: Two-step concentration method by electronegative nitrocellulose membrane and eluted with alkaline glycine/skimmed milk buffer; EM+BE: Two-step concentration method by electronegative nitrocellulose membrane and eluted with alkaline glycine/beef extract buffer; GW+SM: Two-step concentration method by glass wool column and eluted with alkaline glycine/skimmed milk buffer.*

The concentration of high-volume samples (more than 50 L) may require the use of glass wool columns or filters.[17,18,20] However, for many studies on bathing water, well or river water, replicates of 10 L samples may be enough for the study of faecal and/or viral contamination and for MST analysis. In these cases, in order to eliminate the bottleneck associated with two-step methods, a one-step concentration method proposed by Calgua et al. (2008)[21] is used in our laboratory. This method, originally developed for the analysis of sea water, has been optimized so it can also be applied to well and river water,[22,23] and it includes the acidification of a 10L sample. For well or river water, the conductivity of the sample is adjusted to values $\geq$ 1.5 mS/cm^2. The method consists of flocculation by skimmed milk, gravity sedimentation of the flocs and collection of the precipitate to obtain a viral concentrate, which is then dissolved in phosphate buffer, assisted by high speed centrifugation. As shown in Figure 1, this method gives higher viral recovery values with HAdV 2-spiked river and sea water samples than the other two-step concentration methods; (i) electronegative nitrocellulose membrane + elution by beef extract-alkaline glycine buffer, (ii) electronegative nitrocellulose membrane + elution by skimmed milk-alkaline glycine buffer and (iii) glass wool.[23,24] Furthermore, the variability in the viral recovery values given by the two-step concentration procedures was significantly higher than that associated with the direct flocculation method with skimmed milk (coefficient of variation $\leq$ 50%). Calgua et al. (submitted)[23], reported inter-laboratory (Barcelona, Spain and Rio Janeiro, Brazil) values of viral recovery after applying the one-step concentration method for concentrates spiked with HAdV, JCPyV, RV and NoV genogroup II in river water (Figure 2). The viral recovery for all viruses and both laboratories are about 50%, with coefficient of variation values $\leq$ 50%. According to our results and those obtained in collaboration with the laboratory of Marize

Miagostovich (Laboratory of Comparative and Environmental Virology, Oswaldo Cruz Institute, Brazil) in sea and river water after applying the one-step concentration method and specific TaqMan® qPCR for each virus, the procedures represent a low-cost, efficient method to concentrate, detect and quantify viruses routinely in water samples.

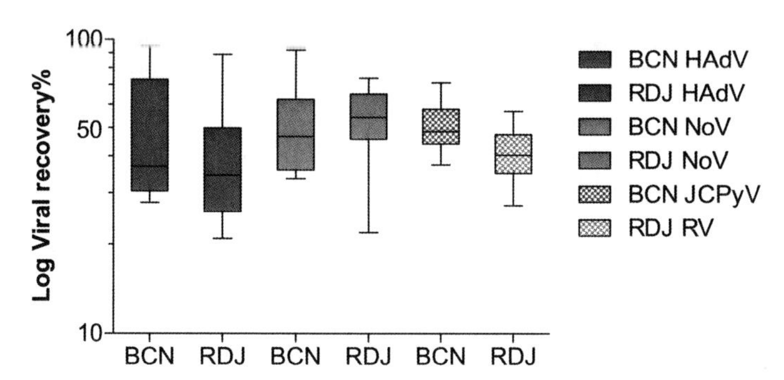

Figure 2 *Inter-laboratory assays Between Barcelona, Spain (BCN) and Rio Janeiro, Brazil(RDJ) to evaluate the viral recovery for spiked DNA and RNA viruses in river water by a one-step concentration method and molecular quantitation by TaqMan®qPCR.*

3 VIRAL INDICATORS AND MICROBIAL SOURCE-TRACKING TOOLS

Classic microbiological indicators such as faecal coliforms, *Escherechia coli* and Enterococci are the indicators most commonly analyzed to evaluate the level of faecal contamination. However, whether these bacteria are suitable indicators of the occurrence and concentration of pathogens such as viruses and protozoan cysts has been questioned for the following reasons: (i) indicator bacteria are more sensitive to inactivation through treatment processes and by sunlight than viral or protozoan pathogens; (ii) the indicator bacteria have non-exclusive faecal source; (iii) some have been shown to multiply in some environments; (iv) the indicator bacteria do not identify the source of faecal contamination; and (v) they have a low correlation with the presence of pathogens.

Various authors concluded that these indicators could fail to predict the risk for waterborne pathogens including viruses.[25,26] Moreover, the levels of bacterial indicators do not always correlate with the concentration of viruses, especially when these indicators are present in low concentrations.[4,27]

The high stability of viruses in the environment, their host-specificity and the high prevalence of some viral infections throughout the year strongly support the use of sensitive molecular techniques for the identification and quantification of the DNA viruses, AdV and PyV, which can be used as indicators of faecal contamination and as MST.

The detection of HAdV in wastewater using molecular methods has been reported widely in the bibliography showing no seasonality.[3,5,6,12,28,29,30,31] The concentration of HAdV in wastewater has been reported to be of up to 3 infectious units/L and of between 10^5-10^8 (GC)/L.[32,33]

HAdV has also been reported in river and surface waters,[4,33,34,45,36,37,38] in sea water,[4,19,21,33,39,40] in swimming pools[41,42] and also in groundwater.[43,44] The presence of HAdV has been reported in drinking water by several authors[35,45,46] and in shellfish.[6,13,30]

JCPyV is ubiquitously distributed and antibodies against JC virus are detected in over 80% of humans worldwide.[47] JCPyV was first suggested as a potential human-specific viral marker in 2000.[5] Since then, the presence of JCPyV alone, or together with BKPyV, under the denomination of HPyV has been reported in several types of environmental samples by PCR-based methods: wastewater from widely divergent geographical areas showing no seasonality,[6,12,31,48] as does river water,[34,49] sea water,[50] drinking water[35,45,51] and shellfish.[6,39] Concentration of JCPyV in wastewater ranged from 10^4 to 10^7 GC/100 mL of wastewater and up to 10^3 GC/100 mL of river water.[33] JCPyV is more frequently excreted than BKPyV,[5] and also than other new human polyomaviruses recently described,[33] is for these reasons that the specific polyomaviral marker in use in our laboratory is based on the quantification of JCPyV.[15] HAdV are commonly from 0.5 to 1 log more concentrated than JCPyV in urban wastewater.[6]

Studies analyzing JCPyV and BKPyV (HPyV) showed a lack of correlation with faecal indicator bacteria[52,53] and their absence in faecal samples of a wide diversity of animals has been reported.[51,53]

The results support the applicability of the proposed indicators as molecular markers of the microbiological quality of water and they would fulfill the conditions required of a human faecal/urine indicator. HAdV and JCPyV may be considered a stable useful marker of human contamination[24] widely disseminated in faecally contaminated environments. It is important to point out, however, that, in some cases, the numbers of specific pathogens in high excretion periods, such as RV or NoV, may exceed the numbers of HAdV.[54]

MST plays a very important role in informing effective management and remediation strategies. MST includes a group of methodologies that aim to identify and, in some cases quantify, the dominant sources of faecal contamination in the environment and, more specifically, in water resources.[55,56] Molecular methods based on molecular detection of host-specific strains of bacteria from the order Bacteroidales and the genus *Bifidobacterium* have been used widely.[57,58] Detection of excreted DNA viruses may allow the development of cost-effective protocols with more accurate quantification of contaminating sources compared to RNA viruses. This is due to the greater accuracy of qPCR and its lower sensitivity to inhibitors, as reverse transcriptase is not used when amplifying DNA viruses. Environmental samples are characterized as complex matrixes and the different variables regarding microbial survival and host specificity have significant impact on the efficacy of all MST approaches. Furthermore, the choice of MST methods and approaches is largely dependent on the objectives of the study considering that the ultimate MST goal is the identification of faecal microbial contamination and its sources in the environment, water and food.

Specific DNA animal viruses have been proposed as MST tools in addition to human viruses, PAdV[7,8,9] and bovine BPyV.[8,10] The use of human and animal viruses analyzed by qPCR as indicators and MST tools has also been discussed and applied by other laboratories.[51,53,56] Harwood et al. (2009)[59] in a study using PCR, suggest that human polyomaviruses were the most specific human marker for MST of the tools analyzed.

In previous studies, we have shown that PAdV and BPyV are widely disseminated in the swine and bovine population respectively, but they do not produce clinically severe diseases. Both viral groups are detected in 100% of the wastewater samples from slaughterhouses tested and also in river water samples near farms. Examples of the concentrations of BPyV and PAdV that have been found in different types of animal and environmental samples by diverse concentration procedures and molecular methods are presented in Table 1.

Table 1 *Results obtained in qPCR assays for BPyV and PAdV in different types of animal and environmental samples.*

Type of sample	BPyV[a]				PAdV[b]			
	n	qPCR		nPCR	*n*	qPCR		nPCR
		Mean	%	%		Mean	%	%
Pig faeces (GC/g)[1]	NT	NT	NT	NT	17	7.25×10^5	100	100[d]
Pig faeces (GC/g)[2]	NT	NT	NT	NT	21	5.58×10^5	76.4	71.4
Slaughterhouse wastewater (GC/L)	11	2.95×10^3	90.9	100[c]	8	1.56×10^6	100	100
River water (GC/L)	6	3.06×10^2	50	83.3	6	8.38×10^0	100	100
Urban wastewater (GC/mL)	8	-	-	-	9	-	-	-
Bovine urine (GC/L)	26	2.21×10^4	30.7	30.7	NT	NT	NT	NT
Bovine faeces (GC/g)	10	-	-	-	NT	NT	NT	NT

[a]*Hundesa et al. (2010)*[10]; [b]*Hundesa et al. (2009)*[9]; ***BPyV:** Bovine polyomavirus;* ***PAdV:** Porcine adenovirus;* ***qPCR:** quantitative PCR;* ***n:** number of samples analyzed;* ***nPCR**: nested PCR;* ***%:** percentage of positive samples;* ***NT:** Non-tested;* ***(-):** Non-detected*

[1]*Pooled pig faeces collected from Catalonia (NE Spain)*

[2]*Pooled pig faeces collected from Basque Country (N Spain)*

[c]*Slaughterhouse wastewater samples analyzed were selected from samples previously found to be positive for BPyV by nPCR*

[d]*Faecal porcine samples analyzed in Catalonia were selected from samples previously found to be positive for PAdV by nPCR*

The results of these studies showed that BPyV and PAdV were quantified in a high percentage (30%-100%) of the samples in which their presence was potentially expected, whereas samples used as negative controls were always negative. BPyV and PAdV were found to be distributed in slaughterhouse wastewater and sludge, and in river water from farm-contaminated areas, but not in urban wastewater collected in areas without agricultural activities.[8,10] These results support the specificity and applicability of the BPyV and PAdV assays for tracing bovine and porcine faecal contamination in environmental samples, respectively.

Several sources of wastewater could contribute to the pollutants found in water bodies. Wastewater treatment plants play an important role in microbiological reduction, minimizing the risks associated with pathogen circulation in the environment, although viruses still pose a challenge to wastewater treatment and disinfection. Hospital wastewater is a particularly interesting matrix, as this commonly contains toxic substances and pathogenic microorganisms.[60,61,62] The effluent wastewater from hospitals is a specific source of contamination of human origin and an interesting matrix in which to analyze the specificity of the proposed MST tools.

Raw wastewater samples from two different hospitals were analyzed to assess the concentration of JCPyV as a human marker in small communities and the specificity of animal (BPyV and PAdV) viral indicators (Table 2). Hospital A (≈100 beds) is located in the area of Girona (Catalonia, NE Spain) and hospital B (>1400 beds) is located in the area of Barcelona (Catalonia, NE Spain). Seven samples from each hospital were collected weekly in a sterile polyethylene container and kept at 4 °C for less than 24h until the virus particles were concentrated. The viral concentration method was based on Pina et al. (1998)[4] and viral DNA was extracted using the QIAmp Viral RNA kit (Qiagen, Inc., Valencia, CA) according to the manufacturer's instructions. The quantification of JCPyV, BPyV and PAdV in the samples was carried out by a TaqMan®-based protocol, as described previously.[8,9,15]

Table 2 *Specificity of the MST tools in human wastewater collected from hospitals.*

Hospital	Sample	JCPyV GC/mL	JCPyV Mean (SD)	BPyV	PAdV
A	1	5.22×10^2	3.15×10^3 (3.34×10^3)	-	-
	2	5.67×10^2		-	-
	3	2.72×10^3		-	-
	4	2.72×10^3		-	-
	5	6.62×10^3		-	-
	6	8.80×10^3		-	-
	7	1.19×10^2		-	-
B	1	4.23×10^2	4.69×10^2 (3.69×10^2)	-	-
	2	3.78×10^2		-	-
	3	3.59×10^2		-	-
	4	7.96×10^2		-	-
	5	1.04×10^2		-	-
	6	1.05×10^2		-	-
	7	1.12×10^3		-	-

Hospital A: One hundred beds, in the area of Girona (Catalonia, NE Spain); **Hospital B:** More than 1400 beds, in the area of Barcelona (Catalonia, NE Spain); **JCPyV:** JC polyomavirus; **BPyV:** Bovine polyomavirus; **PAdV:** Porcine adenovirus; **(-):** Non-detected

The concentration of JCPyV ranged from 10^2 to 10^3 GC/mL in raw wastewater samples from both hospitals and was detected in 100% of the samples analyzed. As expected, BPyV and PAdV were not detected, corroborating the specificity of both viruses for faecal contamination of non-human animal origin. Although hospital wastewater has been described as a source of human pathogenic microorganisms in the environment,[60,61,62] few studies report the presence and concentrations of excreted viruses. Concentrations of HAdV in raw wastewater from hospitals have been reported in Rio de Janeiro (Brasil), ranging around 10^3 GC/mL.[63] To our knowledge, this is the first evaluation and quantification of JCPyV in this type of sample. The concentrations of HAdV and JCPyV measured in wastewater from hospitals are similar to those reported in urban and rural raw wastewater,[24] which indicates that these viruses are consistently excreted by the human population, and can be detected even when small numbers of subjects are studied.

4 IDENTIFICATION OF THE SOURCE OF CONTAMINATION IN GROUNDWATER USING VIRAL MST TOOLS

Groundwater is an important source of drinking water in many regions of the world. Usually, groundwater is considered to have a more stable composition and a higher microbial quality than surface waters. This is partly due to the slow filtration of the water through layers of soil and sediments, which removes pathogenic microorganisms and many chemical compounds.[64] Groundwater is subjected to faecal contamination from a variety of sources, including wastewater treatment plant effluents, land runoff from urban, agricultural, and natural areas, and leachates from sanitary landfills.[65,66] Nevertheless, recent studies have pointed out the role of groundwater as a source of outbreaks in countries of differing economic levels.[67.68,69,70,71] Among waterborne pathogens, enteric viruses have the greatest potential to sink deep into the subsurface environment, penetrate and reach a confined aquifer. The genome of EV, NoV, RV, HAdV or HAV has been detected in many groundwater supplies, including private household wells, municipal wells, and unconfined aquifers.[66,72,73,74,75,76]

Human and animal viruses have been analyzed in groundwater from areas with high levels of nitrates, and used to define the sources of the contamination. Four replicates of ten-liter water samples were collected from 4 different wells in rural areas of Catalonia (Spain). Replicates were collected in each well, and one replicate was seeded with HAdV 35 (kindly provided by Annika Allard, Umeå University, Sweden) and used as process control. Samples were processed according to the protocol previously described by Calgua et al. (2008)[71] and described in detail in Bofill-Mas et al. (2011)[22]. Results obtained are summarized in Tables 3 and 4. The mean values of NO_3 were higher than 100 mg/L in well 1, 2 and 3, whereas the concentration in well 4 was about 30 mg/L. The pH values, conductivity and total dissolved solids were 6.92–7.40, 1205–1894 µS, 7.25–585 ppm, respectively.

Table 3 *Detection of human viruses in groundwater.*

Site	HAdV				JCPyV			
	qPCR	nPCR	N	%	qPCR	nPCR	N	%
1	-	NT	4	0	-	-	5	0
2	-	-	4	0	-	-	5	0
3	-	NT	4	0	-	-	5	0
4	-	-	4	0	-	-	5	0

HAdV: Human adenovirus; **JCPyV:** JC Polyomavirus; **qPCR:** quantitative PCR; **nPCR**: nested PCR; **N:** number of replicates analyzed; **%:** percentage of positive replicates; **NT:** Non-tested; **(-):** Non-detected

Table 4 *Detection of animal viruses in groundwater.*

Site	PAdV				BPyV			
	qPCR	nPCR	N	%	qPCR	nPCR	N	%
1	-	-	4	0	-	-	4	0
2	7.74×10^2	+	5	100	-	-	4	0
3	-	-	4	0	-	-	4	0
4	-	-	4	0	9.53×10^3	+	5	20

qPCR data are expressed as GC/L; **BPyV:** Bovine polyomavirus; **PAdV:** Porcine adenovirus; **qPCR:** quantitative PCR; **nPCR**: nested PCR; **N:** number of replicates analyzed; **%:** percentage of positive replicates; **NT:** Non-tested; **(-):** Non-detected

The virological analysis performed showed the presence of contamination from animal origin in two of the wells. In Well 2 (100% positive samples) contamination of porcine origin was detected, whereas low levels (1/5 positive samples) of bovine contamination were detected in Well 4. In Well 2, the porcine contamination may be related to the presence of porcine slurries in the area, whereas in Well 4, the low level of contamination of bovine origin detected will probably be related to diffuse contamination. No human contamination was present in any of the samples tested. The virological data were confirmed by nested-PCR and sequencing analysis.

In addition to the presence of potential pathogens, discharges of groundwater with its dissolved nitrate into surface water can have an adverse ecological effect, particularly in association with phosphate, leading to eutrophication and causing negative impacts on groundwater-dependent ecosystems.[77,78]

5 CONCLUSIONS

Properly controlling contamination will depend on regulatory authorities choosing and standardizing effective parameters, and then developing surveillance systems with which to monitor and more effectively reduce established, and perhaps prevent emergent diseases. Being able to identify sources of faecal contamination quickly and accurately has important implications for protecting environmental water quality and monitoring for human and potential zoonotic pathogens. Viral indicators may be used not only as an index of viral contamination but also as complementary indicators of faecal/urine contamination in water and food.

The overall conclusions of the studies described above may be summarized as follows:

a. Cost-effective methods for the concentration, detection and quantification of adenoviruses, polyomaviruses and other viral pathogens using molecular methods have been developed and optimized in recent years showing acceptable levels of cost, feasibility, sensitivity and repeatability.

b. The available information supports the proposed human viral indicators, HAdV and JCPyV as useful markers of the presence of human faecal/urine contamination in water and their role as useful tools for quantification of virus removal efficiencies in drinking water and wastewater treatment plants.

c. PAdV and BPyV, are widely excreted in feces and urine respectively and are practical tools for MST in surface and groundwater. The proposed viruses can be quantified by qPCR and are useful for the identification of the biological origin of contamination by nitrates, as well as informing managers on the human, porcine or bovine origin of the contamination.

Acknowledgments

This studies have been partially supported by the Spanish Government "Ministerio de Educación y Ciencia" (project AGL2008-05275-C01/ALI), by the European Union Research Framework 7 funded projects VIROBATHE (Contract No. 513648), VIROCLIME (Contract No. 243923) and by the Catalan Agency of Water, Agència Catalana de l'Aigua (ACA), Departament de Control i Millora dels Ecosistemes Aquàtics. During the developed study Marta Rusiñol was a fellow of the Catalan Government "AGAUR" (FI-DGR).

References

1. S. Pina, A. Creus, N. González, R. Girones, M. Felip, and R. Sommaruga, *J Plankton Res.*, 1998, **20**, 2413. A
2. E.S. Barton, D.W. White, J.S. Cathelyn, K.A. Brett-McClellan, M. Engle, M.S. Diamond, V.L. Miller and H.W. Virgin, *Nature*, 2007, **447**, 326.
3. J. Puig, M. Jofre, F, Lucena, A. Allard, G. Wadell and R. Girones, *Appl. Environ. Microbiol.,* 1994, **60**, 2963.
4. S. Pina, M. Puig, F. Lucena, J. Jofre and R. Girones, *Appl. Environ. Microbiol.*, 1998, **64**, 3376.
5. S. Bofill–Mas, S. Pina and R. Girones, *Appl Environ Microbiol.,* 2000, **66**, 238.
6. S. Bofill-Mas, M. Formiga-Cruz, P. Clemente-Casares, F. Calafell and R. Girones, *J. Virol.*, 2001, **75**, 10290.
7. C. Maluquer de Motes, P. Clemente-Casares, A. Hundesa, M. Martín and R. Girones, *Appl. Environ. Microbiol*, 2003, **70**, 1448.
8. A. Hundesa, C. Maluquer de Motes, S. Bofill-Mas, N. Albinana-Gimenez and R. Girones, *Appl. Environ. Microbiol.*, 2006, **72**, 7886.

9. A. Hundesa, C. Maluquer de Motes, N. Albinana-Gimenez, J. Rodriguez-Manzano, S. Bofill-Mas, E. Suñen and R. Girones, *J. Virol. Methods*, 2009 **158**, 130.
10. A. Hundesa, S. Bofill-Mas, C. Maluquer de Motes, J. Rodriguez-Manzano, A. Bach, M. Casas and R. Girones, *J. Virol. Methods*, 2010, **163**, 385.
11. T. Adrian, G. Schäfer, M.K. Cooney, J.P. Fox and R. Wigand. *Epidemiol. Infect.*, 1988, **101**, 503.
12. S. Bofill-Mas, N. Albiñana-Gimenez, P. Clemente-Casares, A. Hundesa, J. Rodriguez-Manzano, A. Allard, M. Calvo and R. Girones, *Appl. Environ. Microbiol.*, 2006, **72**, 7894.
13. M. Formiga-Cruz, G. Tofiño-Quesada, S. Bofill-Mas, D.N. Lees, K. Henshilwood, A.K. Allard, A-C. Condin-Hansson, B.E. Hernroth, A. Vantarakis, A. Tsibouxi, M. Papapetropoulou, M.D. Furones and R. Girones, *Appl. Environ. Microbiol.*, 2002, **68**, 5990.
14. B.E. Hernroth, A.C. Conden-Hansson, A.S. Rehnstam-Holm, R. Girones and A.K. Allard, *Appl. Environ. Microbiol.,* 2002, **68**, 4523.
15. A. Pal, L. Sirota, T. Maudru, K. Peden and A.M. Lewis, *J. Virol. Methods*, 2006, **135**, 32.
16. P. Clemente-Casares, S. Pina, M. Buti, R. Jardi, M. Martín, S. Bofill-Mas and R. Girones, *Emerg. Infect. Dis.*, 2003, **9**, 448.
17. P. Vilaginès, B. Sarrette, G. Husson, and R. Vilaginès. *Water Sci. Technol.*, 1993, **27**, 299.
18. N. Albinana-Gimenez, P. Clemente-Casares, B. Calgua, S. Courtois, J.M. Huguet and Girones, *J. Virol. Methods.,* 2009, **158**, 104. P. Wyn-Jones, A. Carducci, N. Cook, M. D'Agostino, M. Divizia, J. Fleischer, C. Gantzer, A. Gawler, R. Girones, C. Höller, A. M. de Roda Husman, D. Kay, I. Kozyra, J. López-Pila, M. Muscillo, M. S. Nascimento, G. Papageorgiou, S. Rutjes, J. Sellwood, R. Szewzyk and M. Wyer, *Water Res.*, 2011, **45**, 1025.
19. P. Wyn-Jones, A. Carducci, N. Cook, M. D'Agostino, M. Divizia, J. Fleischer, C. Gantzer, A. Gawler, R. Girones, C. Höller, A. M. de Roda Husman, D. Kay, I. Kozyra, J. López-Pila, M. Muscillo, M. S. Nascimento, G. Papageorgiou, S. Rutjes, J. Sellwood, R. Szewzyk and M. Wyer, *Water Res.*, 2011, **45**, 1025.
20. E. Lambertini, S.K. Spencer, P.D. Bertz, F.J. Loge, B.A. Kieke and M.A. Borchardt, *Appl. Environ. Microbiol.*, 2008, **74**, 2990.
21. B. Calgua, A. Mengewein, A. Grünert, S. Bofill-Mas, P. Clemente-Casares, A. Hundesa, A.P. Wyn-Jones, J.M. López-Pila and R. Girones, *J. Virol. Methods,* 2008, **153**, 79.
22. S. Bofill-Mas, A. Hundesa, B. Calgua, M. Rusiñol, C. Maluquer de Motes and R. Girones, *JoVE*, 2011, In press.
23. B. Calgua, T. Fumian, M. Rusinol, J. Rodriguez-Manzano, S. Bofill, M. Miagostovich and R. Girones, *Submitted for publication.*
24. R. Girones, M.A. Ferrús, J.L. Alonso, J. Rodriguez-Manzano, B. Calgua, A. Corrêa, A. Hundesa, A. Carratala and S. Bofill-Mas, *Water Res.*, 2010, **44**, 4325.
25. C.P. Gerba, S.M. Goyal, R.L. LaBelle, I. Cech, and G.F. Bodgan. *Am. J. Public Health*, 1979, **69**, 1116.
26. E.K. Lipp, S.A. Farrah and J.B. Rose, *Marine Pollution Bulletin*, 2001, **42**, 286.
27. N. Contreras-Coll, F. Lucena, K. Mooijman, A. Havelaar, V. Pierz, M. Boque, A. Gawler, C. Höller, M. Lambiri, G. Mirolo, B. Moreno, M. Niemi, R. Sommer, B. Valentin, A. Wiedenmann, V. Young and J. Jofre, *Water Res.*, 2002, **36**, 4963.
28. M. Formiga-Cruz, A. Hundesa, P. Clemente-Casares, N. Albiñana-Gimenez, A. Allard and R, Girones, *J. Virol. Methods*, 2005, **125**, 111.

29. T.T. Fong, M.S. Phanikumar, I. Xagoraraki and J.B. Rose, *Appl. Environ. Microbiol.*, 2010, **76**, 715.
30. C. Rigotto, M. Victoria, V. Moresco, C.K. Kolesnikovas, A.A. Corrêa, D.S. Souza, M.P. Miagostovich, C.M. Simões and C.R. Barardi, *J. Appl. Microbiol*, 2010, **109**, 1979.
31. P. A. Kokkinos, P. G. Ziros, A. Mpalasopoulou, A. Galanis and A. Vantarakis, *Virol. J.*, 2011, **8**, 195.
32. K.D. Mena and C.P. Gerba, *Ver. Environ. Contam. Toxicol,* 2009, **198**, 133.
33. S. Bofill-Mas, B. Calgua, P. Clemente-Casares, G. La Rosa, M. Iaconelli, M. Muscillo, S. Rutjes, A. M. de Roda Husaman, A. Grunert, I. Gräber, M. Verni, A. Carducci, M. Calvo and P. Wyn-Jones, R. Girones, *Food Environ. Virol.,* 2010, **2**, 101.
34. W. Ahmed, C. Wan, A. Goonetilleke and T. Gardner, *J. Environ. Qual.,* 2010, **39**, 1743.
35. N. Albinana-Gimenez, M. Miagostovich, B. Calgua, J. M. Huguet, L. Matia and R. Girones, *Water Res.*, 2009, **43**, 2011.
36. E. Haramoto, M. Kitajima, H. Katayama, S. Ohgaki, *Water Res.*, 2010, **44**, 1747.
37. L. Ogorzaly, A. Tissier, I. Bertrand, A. Maul and C. Gantzer, *Water Res.*, 2009, **43**, 1257.
38. S. C. Jiang, *Environ. Sci. Technol.,* 2006, **40**, 7132.
39. D.S. Souza, A.P. Ramos, F.F. Nunes, V. Moresco, S. Taniguchi, D.A. Guiguet Leal, S.T. Sasaki, M.C. Bícego, R.C. Montone, M. Durigan, A.L. Teixeira, M.R. Pilotto, N. Delfino, R.M. Franco, C.M. de Melo, A.C. Bainy and C.R. Barardi, *Ecotoxicol. Environ. Saf,* 2011, **76**, 153.
40. Xagoraraki, D.H. Kuo, K. Wong, M. Wong and J.B. Rose, *Appl. Environ. Microbiol.,* 2007, **73**, 7874.
41. M. Papapetropoulou and A.C. Vantarakis, *J. Infect.*, 1998, **36**, 101.
42. J. van Heerden, M.M. Ehlers and W.O. Grabow., *J. Appl. Microbiol.*, 2005, **99**, 1256.
43. L. Ogorzaly, I. Bertrand, M. Paris, A. Maul and C. Gantzer, *Appl. Environ. Microbiol.,* 2010, **76**, 8019.
44. M. Kukkula, P. Arstila, M.L. Klossner, L. Maunula, C.H. Bonsdorff and P. Jaatinen, Scand. *J. Infect. Dis.*, 1997, **29**, 415.
45. N. Albinana-Gimenez, P. Clemente-Casares, S. Bofill-Mas, A. Hundesa and F. Ribas, R. Girones, *Environ. Sci. Technol,* 2006, **40**, 7416.
46. J. van Heerden, M.M. Ehlers, A. Heim and W.O. Grabow, *J Appl Microbiol.*, 2005, **99**, 234.
47. T. Weber, P.E. Klapper, G.M. Cleator, M. Bodemer, W. Lüke, W. Knowles, P. Cinque, A.M. Van Loon, M. Grandien, A.L. Hammarin, M. Ciardi and G. Bogdanovic, *J. Virol. Methods*, 1997, **69**, 231.
48. T.M. Fumian, F.R. Guimares, B.J.P. Vaz, M.T.T. da Silva, F.F. Muylaert, S. Bofill-Mas, R. Girones, J.P.G. Leite and M.P. Miagostovich, *J. Water Health*, 2010, **8**, 438.
49. S. Bofill-Mas, J. Rodriguez-Manzano, B. Calgua, A. Carratala and R. Girones, *Virol. J.*, 2010, **7**, 141.
50. M. Abdelzaher, M. E. Wright, C. Ortega, H. M. Solo-Gabriele, G. Miller, S. Elmir, X. Newman, P. Shih, J. A. Bonilla, T.D. Bonilla, C.J. Palmer, T. Scott, J. Lukasik, V.J. Harwood, S. McQuaig, C. Sinigalliano, M. Gidley, L.R. Plano, X. Zhu, J.D. Wang and L.E. Fleming, *Appl. Environ. Microbiol,* 2010, **76**, 724.
51. S.M. McQuaig, T.M. Scott, J.O. Lukasik, J.H. Paul and V.J. Harwood, *Appl. Environ. Microbiol.,* 2009, **75**, 3379.
52. Korajkic, M.J. Brownell and V.J. Harwood, *J. Appl. Microbiol.*, 2011, **110**, 174.

53. S.M. McQuaig, T.M. Scott, V.J. Harwood, S.R. Farrah and J.O. Lukasik. *Appl. Environ. Microbiol.,* 2006, **72**, 7567.
54. M.P. Miagostovich, F.F. Ferreira, F.R. Guimarães, T.M. Fumian, L. Diniz-Mendes, S.L. Luz, L.A. Silva and J.P. Leite, *Appl. Environ. Microbiol*, 2008, **74**, 375.
55. K.G. Field, *IWA Publishing, London, England*, 2004, 349.
56. T.T. Fong and E.K. Lipp, *Microbiol. Mol. Biol. Rev.*, 2005, **69**, 357.
57. A.E. Bernhard and K.G. Field, *Appl. Environ. Microbiol.*, 2000, **66**, 4571.
58. A.E. Bernhard and K.G. Field, *Appl. Environ. Microbiol.*, 2000, **66**, 1587.
59. V.J. Harwood, M. Brownell, S. Wang, J. Lepo, R.D. Ellender, A. Ajidahun, K.N. Hellein, E. Kennedy, X. Ye and C. Flood, *Water Res*, 2009, **43**, 4812.
60. E. Emmanuel, Y. Perrodin, G. Keck, J.M. Blanchard and P. Vermande, *J. Hazard Mater*, 2005, **117**, 1.
61. E. Emmanuel, M.G. Pierre and Y. Perrodin, *Environ Int.*, 2009, **35**, 718.
62. T. Prado, W.C. Pereira, D.M. Silva, L.M. Seki, A.P. Carvalho and M.D. Asensi, *Lett. Appl. Microbiol.*, 2008, **46**, 136.
63. T. Prado, D.M. Silva, W.C. Guilayn, T.L. Rose, A.M Gaspar and M.P. Miagostovich, *Water Res.*, 2011, **45**, 1287.
64. G. Howard, J. Bartam, S. Pedley, O. Schmoll, I. Chorus and P. Berger, *IWA Publishing, London, England,* 2006, 3.
65. K.J. Schwab, R. De Leon and M.D. Sobsey, *Appl. Environ. Microbiol.*, 1995, **61**, 531.
66. M. Abbaszadegan, P. Stewart and M. LeChevallier, *Appl. Environ. Microbiol.*, 1999, **65**, 444.
67. S.H. Kim, D.S. Cheon, J.H. Kim, D.H. Lee, W.H. Jheong, Y.J. Heo, H.M. Chung, Y. Jee and J.S. Lee, *J. Clin. Microbiol.*, 2005, **43**, 4836.
68. Pedley, S., M. Yates, J.F. Schijven, J. West, G. Howard and M. Barrett. *IWA Publishing, London, England*, 2006, 49.
69. T.T. Fong, L.S. Mansfield, D.L. Wilson, D.J. Schwab, S.L. Molloy and J.B. Rose, *Environ. Health Perspect.*, 2007, **115**, 856.
70. G. Migliorati, V. Prencipe, A. Ripani, C. Di Francesco, C. Casaccia, S. Crudeli, N. Ferri, A. Giovannini, M.M. Marconi, C. Marfoglia, V. Melai, G. Savini, G. Scortichini, P. Semprini and F.M. Ruggeri, *Emerg. Infect. Dis.*, 2008, **14**, 474.
71. G.F. Craun, J.M. Brunkard, J.S. Yoder, V.A Roberts, J. Carpenter, T. Wade, R.L. Calderon, J.M. Roberts, M.J. Beach and S.L. Roy, *Clin Microbiol Rev.*, 2010, **23**, 507.
72. M.A. Borchardt, P.D. Bertz, S.K. Spencer and D.A. Battigelli, *Appl. Environ. Microbiol.*, 2003, **69**, 1172.
73. G.S. Fout, B.C. Martinson, M.W. Moyer and D.R. Dahling, *Appl. Environ. Microbiol.*, 2003, **69**, 3158.
74. M.A. Borchardt, K.R Bradbury, M.B. Gotkowitz, J.A. Cherry and B.L. Parker, *Environ. Sci. Technol.*, 2007, **41**, 6606.
75. J.H. Jung, C.H. Yoo, E.S. Koo, H.M. Kim, Y. Na, W.H. Jheong and Y.S. Jeong, *J. Water Health*, 2011, **9**, 544.
76. L. Guerrero-Latorre, A. Carratala, J. Rodriguez-Manzano, B. Calgua, A. Hundesa and R. Girones, *J. Water Health*, 2011, **9**, 515.
77. D.N. Lerner and R.C. Harris, *Land Use Policy*, 2009, **265**, 265.
78. M.E. Stuart, D.C. Gooddy, J.P. Bloomfield and A.T. Williams, *Sci. Total Environ.*, 2011, **409**, 2859.

DISTINGUISHING POSSUM AND HUMAN FAECES USING FAECAL STEROL ANALYSIS

B.J. Gilpin, M. Devane, D. Wood, A. Chappell

Environmental Health Group, Institute of Environmental Science and Research Ltd, Christchurch, New Zealand

1 INTRODUCTION

The public has become increasingly aware of the potential hazards of faecally contaminated water. This heightened awareness is resulting in an increased frequency of water quality monitoring for the traditional microbial indicators, faecal coliforms, *Escherichia coli* and enterococci. There is also an expectation that when these indicators are detected, corrective action will be taken to eliminate these faecal indicators - and by inference the faecal pollution - from the water. While these traditional indicators are usually a good indication of microbial quality, and therefore the risk posed, they provide little guidance as to the source of the faecal pollution. Faecal coliforms and other traditional indicators are present in the faeces of humans, cows, sheep, dogs, ducks, seagulls and a wide range of other animals. Identifying the source of faecal pollution can be crucial for effective water management.

A range of methods have been developed to identify sources of faecal contamination including PCR analysis of specific DNA fragments, culturing of alternative organisms, and detection of chemicals including fluorescent whitening agents (FWAs), caffeine, and faecal sterols[1]. One key step in the validation of each method is to evaluate the presence of each marker in different animal sources. In New Zealand efforts have primarily focussed on faecal outputs from the human population (4.5 million), dairy cows (5.3 million), beef cattle (4.4 million), sheep (32 million) and various populations of wildfowl[2, 3].

One potential source of water faecal pollution that has not been widely considered is the brushtail possum. The common brushtail possum (*Trichosurus vulpecula*) is a small marsupial that typically weighs between 1.5-3.0 kg. It was introduced into New Zealand from Australia by Europeans in the 1800s to establish a fur industry. After its introduction, the possum became established as one of New Zealand's most serious mammalian pests, with numbers currently estimated at 70 million[4]. Daily faecal outputs of possums have been estimated at 25.2 g dry weight of faeces per day[5], which corresponds to a potential total daily faecal output of over 1.5 million kg dry weight. Whilst much of this deposition occurs in forest areas, possum densities have been estimated at 10-12 per hectare in scrub-filled swamps, and 5 per hectare in streamside willows[6].

In this study, we evaluated whether faecal sterol analysis could be used to identify faecal pollution from possums. Faecal sterols are a group of C27-, C28- and C29- cholestane-based sterols found mainly in animal faeces. The sterol profile of faeces depends on the interaction of three factors. First, the animal's diet determines the relative quantities of sterol precursors (cholesterol, 24-ethylcholesterol, 24-methylcholesterol, and/or stigmasterol) entering the digestive system. Second, animals differ in their endogenous biosynthesis of sterols (for example, human beings on a low cholesterol diet synthesise cholesterol). Third, and perhaps most important, is that the anaerobic bacteria in the animal gut biohydrogenate sterols to stanols of various isomeric configurations[7, 8]. The sterol, cholesterol, can be hydrogenated to one or more of four possible stanols. In human beings, cholesterol is preferentially reduced to coprostanol, whereas in the environment cholesterol is predominately reduced to cholestanol. Similarly, plant-derived 24-ethylcholesterol is reduced to 24-ethylcoprostanol and 24-ethylepicoprostanol in the gut of herbivores, whereas in the environment it is primarily reduced to 24-ethylcholestanol.

We examined the levels of sterols in possum faeces, and in comparison with faeces from other animals developed criteria for distinguishing possum faeces, from that of other sources.

2 METHODS AND RESULTS

2.1 Sterol Analysis of Possum Faeces

Faecal sterols were extracted directly from faecal samples (0.03g), which were spiked with a deuterated internal standard and refluxed with 6% methanolic KOH for 4 hours. The supernatant containing the hydrolysed sterols was cleaned by centrifugation, the sterols were partitioned into hexane and then evaporated to dryness. Each sample was derivatised, a system monitoring compound added, and analysed by gas chromatography with mass spectrometric detection[9].

Analysis of 10 possum faecal samples identified an average of 1,568 nanograms of total sterols/gram of faeces. The mean levels of each sterol detected, with 95% confidence levels are given in Table 1.

Sterol profiles from possums were then compared with sterol profiles obtained from 28 human samples (8 human septic tank samples, 20 human wastewater samples), 42 herbivore samples (10 sheep faeces, 8 rabbit faeces, 1 horse, 5 dairy shed effluent, 5 beef works effluent, 13 sheep works effluent), 19 avian samples (10 duck faeces, 7 seagull faeces, 2 chicken faeces). Extraction of these samples was performed in the same manner, except for effluent samples which were first filtered onto glass fibre filters (up to 1 L), before continuing as described above. Comparisons were performed on the basis of percentage of each sterol present (Figure 1).

Table 1. *Average sterols measured in nanograms per gram of possum faeces.*

Sterol	Mean (95% Confidence Interval)
Coprostanol	177 (133 to 221)
24-ethylcoprostanol	699 (374 to 1,024)
Epicoprostanol	8 (2 to 14)
Cholesterol	179 (47 to 311)
Cholestanol	11 (7 to 15)
24-methylcholesterol	82 (64 to 100)
24-ethylepicoprostanol	28 (3 to 53)
Stigmasterol	21 (18 to 24)
24-ethylcholesterol	237 (202 to 271)
24-ethylcholestanol	126 (101 to 151)
Total sterols	1,568 (1,074 to 2,063)

Possum faeces had a number of potentially distinctive proportions of sterols. In particular, the level of 24-ethylcoprostanol for possums is higher than the other groups and that the level of cholesterol is lower for possums than other groups.

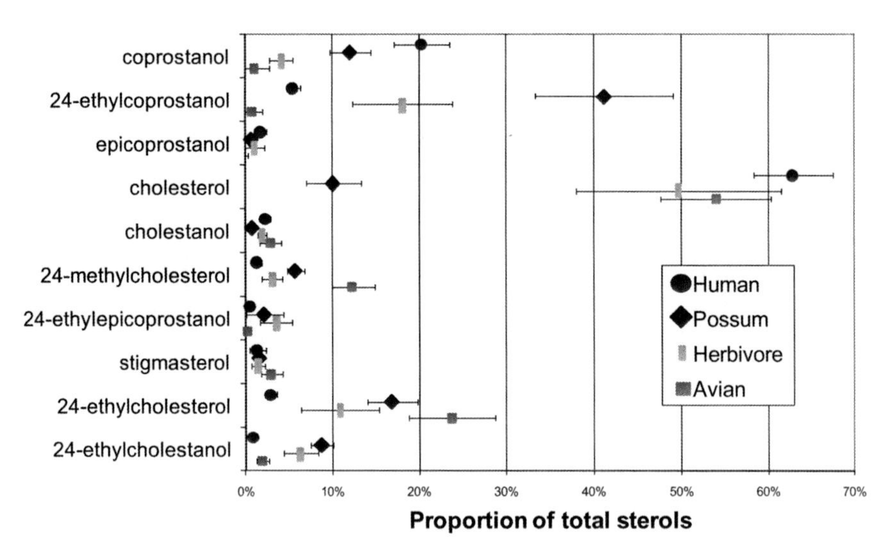

Figure 1 *Comparison between the average normalized levels of sterols in samples by types. Error bars indicate the 95% confidence interval around the mean.*

Two ratios were identified (coprostanol/cholestanol > 6.0 and coprostanol/24-ethylcoprostanol < 1.0), which when applied to the 99 items in the data set, correctly classified the possums (Figure 2).

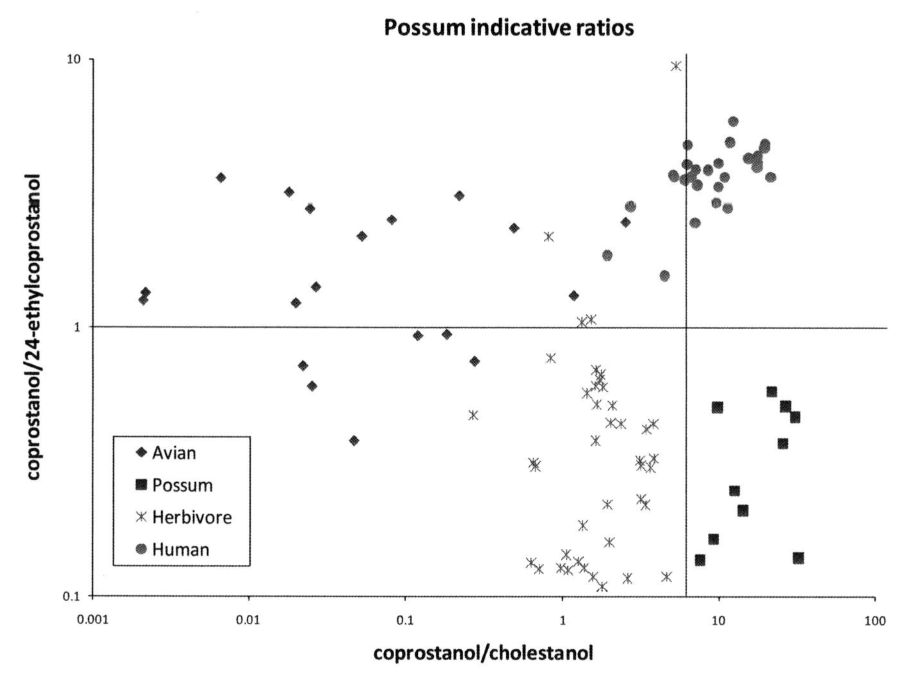

Figure 2 *Evaluation of sterol ratios from possum faeces and a range of other faeces from human, herbivore and avian sources.*

To test the validity of these ratios in environmental samples, three freshly collected possum faecal samples were spiked into river water samples (0.1 g/100ml). One hundred ml of each was filtered, and extracted as described above. Each of the spiked samples contained elevated levels of at least five of the sterols (Table 2). The ratio of coprostanol/cholestanol was greater than 6.0 in these spiked samples and, together with the ratio of coprostanol/24-ethylcoprostanol being less than 1.0, possum faeces would have been identified in these spiked river samples.

For comparison, thirty four river samples from around New Zealand were also analysed, including several suspected to be contaminated with human, ruminant or avian faeces. None of these samples were consistent with possum faecal contamination. Three examples of these naturally contaminated river water samples are given in Table 2.

Table 2. *Faecal sterol analysis of river water samples.*

Sterols	River water	River water spiked with Possum A faeces	Possum B faeces	Possum C faeces	River water samples with suspected contamination with Human faeces	Avian faeces	Bovine faeces
coprostanol	3	135	83	116	177	60	2628
24-ethylcoprostanol	4	605	317	675	84	296	7463
epicoprostanol	1	2	2	2	4	20	543
cholesterol	155	414	375	355	441	4210	2833
cholestanol	7	12	9	12	44	383	1302
24-methylcholesterol	198	217	403	347	98	957	955
24-ethylepicoprostanol	1	1	4	3	2	40	3200
stigmasterol	1101	1847	2082	2020	78	241	216
24-ethylcholesterol	1413	2200	4093	3186	198	2021	3302
24-ethylcholestanol	10	19	14	25	13	179	10459
Total Sterols	2893	5452	7382	6741	1139	8407	32901
Ratios							
coprostanol/cholestanol	0.4	11.3	9.2	9.7	4.0	0.2	2.0
coprostanol/24-ethylcoprostanol	0.8	0.2	0.3	0.2	2.1	0.2	0.4

3 CONCLUSION

Possum faeces contain significant levels of sterols. However there are a number of distinctive features of the sterol profile of possums which could be used to distinguish possum faeces from other sources. Similar to human faeces, possums contain high levels of coprostanol which distinguishes them from herbivores such as sheep and cattle. However in contrast to human faeces, possum faeces contain high levels of 24-ethylcoprostanol. Ratios based on these differences (coprostanol/cholestanol > 6.0 and coprostanol/24-ethylcoprostanol < 1), when both present, appear indicative of possum faeces. Based on a survey of river water samples, possum faeces was not a significant contamination source in New Zealand. Source determinations should be supported by other faecal source tracking tools and consideration of potential sources actually present in a catchment.

References

1. K. G. Field and M. Samadpour, *Water Res.*, 2007, **41**, 3517-3538.
2. B. J. Gilpin, B. Robson, P. Scholes, F. Nourozi and L. W. Sinton, *Lett. Appl. Microbiol.*, 2009, **48**, 162-166.
3. E. M. Moriarty, M. L. Mackenzie, N. Karki and L. W. Sinton, *Appl. Environ. Microbiol.*, 2011, **77**, 1797-1803.
4. A. McDowell and B. J. McLeod, *Adv Drug Deliv Rev*, 2007, **59**, 1121-1132.
5. A. Fitzgerald, .E., *Proceedings of the New Zealand Ecological Society*, 1977, **24**, 76-78.
6. P. E. Cowan, in *The Handbook of New Zealand Mammals* ed. C. M. King, Oxford University Press, Auckland, Editon edn., 1990, pp. 68-98.
7. M. Devane, D. Saunders and B. Gilpin, *Chem. N.Z.*, 2006, **70**, 74-77.
8. R. Leeming, N. Bate, R. Hewlett and P. D. Nichols, *Water Sci. Technol.*, 1998, **38**, 15-22.
9. S. M. Mudge and C. E. Norris, *Mar. Chem.*, 1997, **57**, 61.

RAPID CONFIRMATION OF PRESUMPTIVE *Clostridium perfringens* COLONIES BY POLYMERASE-CHAIN REACTION.

R. Múrtula, E. Soria, M.A. Yáñez and V. Catalán.

LABAQUA, S.A. Pol Ind. Atalayas, C/ Del Dracma 16-18 03114-Alicante, Spain.
E-mail: vicente.catalan@labaqua.com

1 INTRODUCTION

Clostridium perfringens is considered an important index microorganism of water pollution and a useful marker to alert water companies to the possible presence of other stress-resistant pathogens.[1] In addition, its resistance to chlorination is useful as an indicator of the effectiveness of water treatment processes. Levels of *C. perfringens* in raw water are commonly monitored to detect an increase in the risk of microbiological contamination.[2] Furthermore, the presence of *C. perfringens* in drinking water is considered significant and warrants immediate investigation. In Europe, *C. perfringens* has been used in conjunction with other sulphite reducing clostridia to monitor faecal contamination in water since 1960s. [3]

The standard methodology used for the detection of *C. perfringens* in water is primarily based on bacterial growth on specific culture media and subsequent confirmation by a series of biochemical tests.[4,5] However, these standard microbiological and biochemical methods present a number of disadvantages: *i*) the time needed to obtain results is usually long since it depends on the growth period of the microorganism, *ii*) confirmation tests frequently produce ambiguous results, and *iii*) certain tests present low sensitivity and specificity.[6,7]

The available culture media for *C. perfringens* are based on the use of defined substrates and commercially correspond to the chromogenic medium membrane *Clostridium perfrigens* (m-CP) agar and Perfringens base agar with selective supplement Tryptose sulphite cycloserine (TSC) for *C. perfringens*. Compared to traditional methods, m-CP medium provides faster results with high selectivity and specificity, although its low recovery rate makes TSC media more useful than m-CP.[8] However, TSC requires the use of biochemical tests for presumptive colonies confirmation and at least the study of motility, nitrate reduction, lactose fermentation, and gelatin liquefying.[5] These confirmation methods present a number of disadvantages such as long time-to-results, and frequent ambiguous results. Diagnostic microbiology laboratories need more rapid and reliable analytical tools that improve the use of traditionally employed methodologies.

In this sense, nucleic acid-based technologies, specially, the amplification of nucleic acids using polymerase chain reaction (PCR), have resulted in faster and easier bacterial detection.[9,10] Hence, a great number of rapid methods for the detection of important pathogenic bacteria in both water and food samples employing PCR have been developed.[11] The common characteristics of all PCR methodologies are increased specificity, sensitivity, speed, and low limit of detection. In spite of these advantages, many PCR protocols are not easily applicable to routine bacterial analysis because many diagnostic laboratories lack the technology or time for the full development, validation, and implementation of such methods. Indeed, a lack of normalization and regulation together with frequent heterogeneous results has lead to the classification of these methods as mere scientific approaches rather than routine diagnostic tools. Moreover, additional disadvantages, such as the presence of PCR inhibitors, a lack of automation (e.g., for sample preparation), and difficulty in quantifying living cells, remain. Perhaps due to the above shortcomings, most of the current legislation on microbiological quality testing only includes conventional culture isolation methodologies. Therefore, in spite of the benefits of PCR, culture isolation remains a compulsory tool in the majority of diagnostic microbiology laboratories.

While PCR technologies are evolving and their current drawbacks are being resolved, in this work we have developed a conventional PCR confirmation system for confirming the identity of presumptive *C. perfringens* colonies that permits water-testing laboratories, using conventional TSC culture isolation methods, to improve the confirmation steps, avoiding their usual drawbacks. Moreover, this methodology permits the simultaneous confirmation of different species of presumptive isolated colonies using the same thermal cycler program. The final objective of this work is to extend the simultaneous PCR confirmation method to a set of 10 additional bacteria, including *Acinetobacter baumanii*, *Escherichia coli* O157:H7, *Campylobacter jejuni*, *Listeria monocytogenes*, *Pseudomonas aeruginosa*, *Salmonella* spp., *Staphylococcus aureus*, *Vibrio cholerae*, and *Yersinia enterocolitica.*

2 MATERIALS AND METHODS

2.1 Bacterial strains

All the strains used for validation purposes were cultivated according to their corresponding Type Culture Collection instructions. A total of 100 strains from 19 different genera were tested (Table 1); fifty-three of the strains were from culture collection strains and the remainder were environmental and clinical isolates (e.g., *Staphylococcus* spp., *Enterococcus* spp., *Klebsiella* spp., *Pseudomonas* spp., *Escherichia* spp., *Citrobacter* spp., *Aeromonas* spp., *Listeria* spp., *Yersinia* spp., *Shigella* spp., *Vibrio* spp., *Proteus* spp., *Serratia* spp., *Hafnia* spp., *Bacillus* spp., *Micrococcus* spp., *Helicobacter* spp., *Campylobacter* spp., and *Acinetobacter* spp.). In this study, 15 *C. perfringens*, and 8 *Clostridium* spp. environmental strains were also included.

2.2 DNA extraction and PCR amplification

Bacterial colonies were resuspended in 200 µL of 20% Chelex-100 resin (Bio-Rad Laboratories, Richmond, CA). Total genomic DNA was then extracted by three freeze-thaw cycles (-75ºC for 10 min and 94ºC for 10 min), and cellular debris were removed by centrifugation at 10,000 xg for 1 min. Finally, the purity of the obtained DNA was evaluated by spectrophotometer measures at 260/280 nm in a Biophotometer (Eppendorf, Hamburg, Germany).

Table 1 *Strains used for testing the specificity of the PCR assay.*

Organism[a]	Source	Organism[a]	Source
Acinetobacter baumanii	NCTC 12156	*Pseudomonas aeruginosa*	ATCC 27865
Aeromonas hydrophila	CECT 839	*Pseudomonas aeruginosa* V227	Environment
Bacillus subtilis	CECT 4522	*Pseudomonas aeruginosa* V313	Environment
Campylobacter jejuni	NCTC 11351	*Pseudomonas aeruginosa* V314	Environment
Campylobacter fetus	NCTC 10842	*Pseudomonas aeruginosa* V315	Environment
Campylobacter coli	NCTC 11366	*Pseudomonas* sp. V215	Environment
Campylobacter lari	NCTC 11352	*Pseudomonas* sp. V226	Environment
Citrobacter freundii	CECT 401	*Pseudomonas* sp. V236	Environment
Clostridium bifermentans	CECT 550	*Pseudomonas* sp. V233	Environment
Clostridium perfringens	CECT 376	*Pseudomonas* sp. V234	Environment
Enterococcus faecium	CECT 184	*Pseudomonas* sp. V243	Environment
Enterococcus faecium	CECT 410	*Pseudomonas marginalis*	CECT 229
Enterococcus faecium	NCTC 7380	*Pseudomonas putida*	CECT 324
Enterococcus faecium	NCTC 8619	*Pseudomonas stutzeri*	CECT 930
Enterococcus faecium	NCTC 12201	*Pseudomonas alcaligenes*	NCTC 10367
Enterococcus faecium	NCTC 12202	*Salmonella enterica*	CECT 409
Enterococcus faecium	NCTC 12203	*Salmonella* sp. 192626	Environment
Enterococcus faecalis	DSM 20376	*Serratia marcescens*	CECT 846
Enterococcus faecalis	CECT 4176	*Shigella boydii*	CECT 583
Enterococcus faecalis	NCTC 8176	*Shigella dysenteriae*	NCTC 9934
Enterococcus faecalis	CECT 407	*Shigella dysenteriae*	NCTC 8599
Enterococcus faecalis	CECT 408	*Shigella dysenteriae*	NCTC 11869
Enterococcus durans	CECT 411	*Shigella dysenteriae*	NCTC 9953
E. coli O157:H7	ATCC 43888	*Shigella flexneri*	CECT 585
Escherichia coli	CECT 434	*Shigella sonnei*	CECT 413
Escherichia coli	ATCC 35329	*Shigella sonnei*	CECT 457
Escherichia coli	ATCC 35354	*Staphylococcus aureus*	CECT 86
Escherichia coli	ATCC 35363	*Staphylococcus aureus*	CECT 239
Escherichia coli	ATCC 35368	*Staphylococcus aureus*	CECT 435
Escherichia coli	ATCC 35357	*Staphylococcus aureus*	CECT 438
Escherichia coli	ATCC 35377	*Staphylococcus aureus*	CECT 520
Hafnia alvei	CECT 157	*Staphylococcus aureus*	CECT 826
Helicobacter pylori	NCTC 11637	*Staphylococcus aureus*	CECT 794
Helicobacter pylori	NCTC 12489	*Staphylococcus aureus*	CECT 827
Listeria monocytogenes	CECT 4031	*Staphylococcus aureus*	CECT 957
Listeria innocua	CECT 910	*Staphylococcus aureus* V279	Environment
Listeria ivanovii	CECT 913	*Staphylococcus aureus* V284	Environment
Listeria seeligerii	CECT 917	*Staphylococcus aureus* V303	Environment
Listeria welshimeri	CECT 919	*Staphylococcus hyicus*	NCTC 1035
Listeria grayi	CECT 931	*Staphylococcus intermedius*	NCTC 11048
Micrococcus sp. V182	Environment	*Vibrio cholerae*	CECT 512
Micrococcus luteus	CECT 4057	*Vibrio cholerae*	CECT 552
Proteus vulgaris	CECT 165	*Yersinia enterocolitica*	NCTC 10598
Pseudomonas aeruginosa	CECT 108	*Yersinia enterocolitica*	NCTC 11177
Pseudomonas aeruginosa	CECT 109	*Yersinia enterocolitica*	CECT 4054
Pseudomonas aeruginosa	CECT 116	*Yersinia enterocolitica*	CECT 4315
Pseudomonas aeruginosa	CECT 118	*Yersinia enterocolitica*	CECT 500
Pseudomonas aeruginosa	CECT 519	*Yersinia enterocolitica*	NCTC 11600
Pseudomonas aeruginosa	CECT 4400	*Yersinia enterocolitica*	CETC 4055
Pseudomonas aeruginosa	CECT 4092	*Yersinia enterocolitica*	NCTC 11599

[a]CECT, Spanish Type Culture Collection, Valencia, Spain; NCTC, National Collection of Type Cultures, Central Public Health Laboratory, London, United Kingdom; ATCC, American Type Culture Collection, Manassa, VA, USA; DSM, Deutsche Sammlung von Mikroorganismen und Zellkulturen GmbH, Braunschweig, Germany.

A pair of previously described oligonucleotide primers were selected to confirm by PCR presumptive colonies of *C. perfringens* (Table 2) and were synthesized externally (Sigma Aldrich, St. Louis, MO). An experimental confirmation of the specificity of the selected primer set was performed using genomic DNA from all the bacterial strains described before.

Table 2 *Selected primer set for the conventional* C. perfringens *PCR confirmation.*

Target	Primer	Sequence (5'→3')	Size of the PCR product (bp)	Reference
Phospholipase C	PhF	AAGTTACCTTTGCTGCATAATCCC	290	Fach and Gillou, 1993
	PhR	ATAGATACTCCATATCATCCTGCT		

Each 40-μL reaction mixture for the detection of *C. perfringens* consisted of 2 μL of template DNA and 38 μL of Master mix containing 1 U of Taq polymerase (Eppendorf, Hamburg, Germany), 200 nM of each primer (Sigma Aldrich, St. Louis, MO), 0.2 nM of dNTPS (Roche, Mannheim, Germany), $MgCl_2$ 1.5 mM, 1X PCR buffer (Eppendorf, Hamburg) and an Internal Positive Control (IPC) DNA. The IPC was constructed in order to be amplified with the same primer set than the target but always yielding an amplification product with higher size (650 bp) and their construction was performed as previously described.[12]

PCR amplifications were performed in a GeneAmp PCR System 9700 thermal cycler (Applied Biosystems, Foster City, CA) and the amplification conditions were the following: heat denaturation at 94°C for 4 min followed by 35 cycles of heat denaturation at 94°C for 30 s, primer annealing at 62°C for 30 s and DNA extension at 72°C for 1 min, followed by a final extension step of 10 min at 72°C. A synthetic plasmidic DNA containing the gene target was used as positive control, and sterile water was used as negative control for detecting possible reagents contamination. PCR-products were analyzed by gel electrophoresis in 1% agarose (Pronadisa, Madrid, Spain) containing 0.5μg/mL ethidium bromide in Tris-borate buffer.[13] DNA bands were visualized by UV illumination and photographed using the Gel Doc 2000 documentation system (Bio-Rad Laboratories, Richmond, CA).

2.3 Analysis of natural samples

In order to evaluate the feasibility of the method, a total of 341 *C. perfringens* presumptive positive samples (67 continental waters and 274 wastewaters) were analyzed. Following the ISO 6461-2, 100-mL of water were concentrated by filtration through cellulose membranes (0.4-μm pore size and 47-mm diameter) and cultured in TSC agar under anaerobic conditions at 37°C for 24 hours. Five presumptive colonies for each analyzed sample were confirmed in parallel by PCR and the conventional biochemical confirmation tests, using the MNLG method[5] (motility, nitrate reduction, lactose fermentation, and gelatin liquefying). To identify the colonies with ambiguous results (differences between biochemical and PCR confirmation), 16S rDNA sequencing was performed as previously described.[14]

3 RESULTS

3.1 PCR design

Most of the published PCR designs are based on genes that only amplify pathogenic *Clostridium*[15,16,17], and it is difficult to find a primer set that amplifies all *C. perfringens.* Furthermore, in most cases, an internal positive control (IPC) is rarely included, even though its use should be considered compulsory.[18]

After considering all previously published conventional amplification systems [16, 17, 19, 20, 21, 22] the *C. perfringens* alpha-toxin gene (phospholipase C) was selected. The set of primers based on this gene showed to be specific for all *C. perfringens* (both pathogens and non-pathogens) and had demonstrated good specificity in a previous study.[21]

3.2 Specificity study

The theoretical specificity of the primers was assessed using the BLAST family program package.[23] In addition to this bioinformatics approach, an experimental confirmation of the specificity of the primer set was tested with the bacterial strains previously mentioned in Material and Methods. After amplification, the selected PCR system yielded the expected PCR products for *C. perfringens* strains; and, no other bacteria yielded a positive signal.

3.3 Performance of the assay

From the 341 presumptive water samples tested in order to evaluate the feasibility of the method, a total of 1,705 *C. perfringens* colonies were confirmed. Of the tested colonies, 38.1% (650) were negative by both methods (PCR and biochemical confirmation tests), and 60.5 % (1,032) were positive by both methods. These results indicate that the biochemical tests were in agreement with the PCR results in 98.6% of the tested colonies. However, 23 colonies (1.4%) were positive using the biochemical tests and negative by PCR. These isolates with discordant results were identified as *C. sporogenes* by 16S rDNA sequencing.

Figure 1 shows an example of a confirmation test for *C. perfringens* in a continental water sample.

In addition, the analysis of data from the 3rd round of the drinking water ielab proficiency testing scheme (PTS) in 2010, in which 168 laboratories participated, showed 14% of false positives when the sample was contaminated only with *C. sporogenes* and confirmed with biochemical tests.

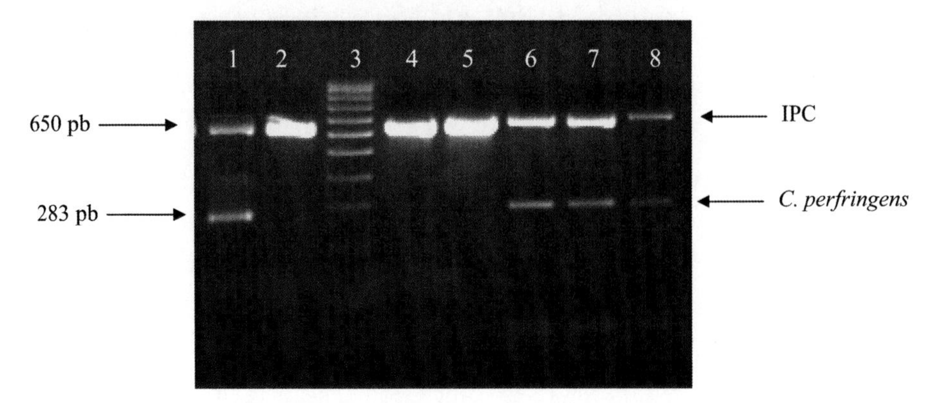

Figure 1. *Agarose electrophoresis gel showing a PCR confirmation assay of Clostridium perfringens from a continental water sample.* Lane 1, positive control for *C. perfringens*; Lane 2, negative control for *C. perfringens*; Lane 3, molecular weight marker VIII (Roche); Lanes 4 and 5, presumptive colony not confirmed as *C. perfringens*; Lanes 6-8, presumptive colonies confirmed as *C. perfringens*.

4 DISCUSSION

In an attempt to accelerate the confirmation stage, conventional PCR using the previously described gene-specific primers for *C. perfringens* was evaluated. For this, 1,705 colonies of presumptive *C. perfringens* from environmental samples were confirmed in parallel by PCR and conventional biochemical protocols.[5]

One of the main advantages of this system is the possibility of simultaneous detection of several microorganisms (e.g. different faecal index in water). Sample preparation, once the presumptive colonies are obtained, and the thermal cycler program, using specific primers for each target microorganism, is common for all the mentioned organisms. Remarkably, this approach does not lead to any difficulties in interpretation since an IPC is included to exclude false-negative results from PCR inhibition. In addition, the easy interpretation of results, as well as the speed of this procedure, will facilitate the implementation of this system in routine diagnostic laboratories. This system allows confirming the identity of presumptive colonies of *C. perfringens* (or any other target microorganism) in less than three hours once the DNA is obtained, versus the several days that are often required for conventional biochemical methods. The approach presented in this work, exploits many of the advantages of PCR (e.g., high sensitivity, specificity, speed, and throughput).

On the other hand, the three main disadvantages of conventional PCR, that is *i*) false-negative results due to PCR inhibition, *ii*) the qualitative nature of the assay, and *iii*) the difficulty in distinguishing between viable and non-viable microorganisms, are not relevant in this particular application. The first drawback is negligible because of the use of an IPC in our assay and the last two disadvantages do not affect the usefulness and feasibility of the method since it is initiated from isolated colonies as a confirmative test.

It is also important to mention that certain cost-related features must be added to the advantages of this simultaneous PCR system. This method avoids the use of expensive reagents for biochemical testing, as well as for the DNA extraction step. Moreover, the speed of this method allows same-day results, thus saving time and money.

This simple method can be implemented by most water- and food-testing laboratories as an alternative to classical methodologies, making the confirmation step faster, easier, and more specific. In any case, the development of more advanced techniques that allow the simultaneous and completely automatized detection of living microorganisms, such as microarrays and so-called "lab-on-a-chips" continues. Meanwhile, conventional PCR systems, such as the one proposed in this work, can be considered an initial step in the implementation of more powerful molecular detection tools that can play important roles in routine diagnostic microbiology to provide rapid, accurate, and cost-effective results when conventional biochemical methods are not available, or rapid and specific enough.

References

1 J.W. Bisson, and V.J. Cabelli, *J. Water Pollut. Con. F,* 1980, **52**.
2 P. Payment, and E. Franco, *Appl. Environ. Microbiol.*, 1993, **59,** 2418.
3 Anonymous, European Council Directive 98/83/EC on the quality of water intented for human consumption, *European Commission Environment,* 1998.
4 Anonymous, Water quality-Detection and enumeration of the spores of sulphite-reducing anaerobes (Clostridia) Part 2: Method by membrane filtration (ISO 6461-2) *European Committee for standardization,* 1986.
5 Anonymous, Microbiology of food and animal feeding stuffs - Horizontal method for the enumeration of *Clostridium perfringens* - Colony-count technique (ISO 7937:2004), *European Committee for standardization,* 2004.
6 C.W. Blackburn, *J. Appl. Bacteriol.,* 1993, **75**, 199.
7 B. Swaminathan, and P. Feng, *Annu. Rev. Microbiol.*, 1994, **48**, 401.
8 M. Araujo, R. A Sueiro, M. J. Gómez, and M. J. Garrido, *J. Microbiol. Meth.,* 2004, 175.
9 P.T. Monis, and S. Giglio, *Infect. Genet. Evol.*, 2006, **6**, 2.
10 R. Naravaneni, R. K. Jamil, *J. Med. Microbiol.*, 2005, **54**, 51.
11 S. Vollenhofer-Schrumpf, R. Buresch, G. Unger, N. Stahl, G. Fränzl, and M. Schinkinger, *J. Rapid Meth. Autom. Microbiol.*, 2005, **13**, 148.
12 M.A. Yáñez, C. Carrasco-Serrano, V.M. Barberá, and V. Catalán, *Appl. Environ. Microbiol.*, 2005, **71**, 3433.
13 J. Sambrook, E.F. Fritsch, and T. Maniatis, T. in *Molecular cloning. a laboratory manual,* 2nd edition. Cold Spring Harbor Laboratory Press, New York, USA. 1989.
14 M.A Yáñez, V. Catalán, D. Apráiz, M.J. Figueras, and A.J. Martínez-Murcia, Int. J. Syst. Evol. Microbiol., 2003, **53**, 875.
15 A. Heikinheimo, M. Lindström, and H. Korkeala, *J. Clin. Microbiol.*, 2004, **42**, 3992.
16 C. Herholz, R. Miserez, J. Nicolet, J. Frey, M. Popoff, M. Gibert, H. Gerber, and R. Straub, *J. Clin. Microbiol.,* 1999, **37**, 358.
17 M. Saito, M. Matsumoto, and M. Funabashi, *Int. J. Food Microbiol.,* 1992, **17**, 47.
18 J. Hoorfar, N. Cook, B. Malorny, M. Wagner, D. De Medici, A. Abdulmawjood, and P. Fach, *J. Appl. Microbiol.,* 2004, **96**, 221.
19 S. Aabo, O.F. Rasmussen, L. Rossen, P.D. Sørensen, and J.E. Olsen, *Mol. Cell probes,* 1993, **7**, 171.
20 E. Augustynowicz, A. Gzyl, and J. Slusarczyk, *J. Med. Microbiol.,* 2002, **5**, 169.
21 P. Fach, and J.P. Guillou, *J. Appl. Bacteriol.,* 1993, **74**, 61.
22 M.N. Widjojoatmodjo, A.C. Fluit, R. Torensma, B.H. Keller, and J. Verhoef, *Eur. J. Clin. Microbiol. Infect. Dis.,* 1991, **10**, 935.
23 S.F. Altschul, W. Gish, W. Miller, E.W. Myers, and D.J. Lipman, *J. Mol. Biol.,* 1990, **215**, 403.

AN EVALUATION OF BACTERIAL SOURCE TRACKING OF FAECAL BATHING WATER POLLUTION IN THE KINGSBRIDGE ESTUARY, UK

K. R. Hussein[1], G. Bradley[1]and G. Glegg[2]

[1]School of Biomedical and Biological Sciences, University of Plymouth, PL4 8AA, UK.
[2] School of Marine Sciences and Engineering, University of Plymouth, PL4 8AA, UK.

1 INTRODUCTION

Vinten *et al.* [1] referred to certain areas of the UK such as South Wales, North Yorkshire and South West Scotland which have problems with bathing water compliance. Areas of intensive dairy farming with a high cattle concentration, cool, humid summers are prone to bathing water contamination [2]. Farm areas and cattle walkways are highly susceptible to direct runoff of microbial contaminated water to streams, and there is widespread direct use of streams to supply drinking water to stock in summer. Heavy rainfall can exacerbate these inputs from agricultural land and can also cause problems with sewage treatment systems, which if overloaded, may resort to the use of storm sewers overflows, discharging sewage and rainfall largely untreated[3]. Contamination of water systems can cause a higher risk to bathers, as well as economic losses as a result of closed beaches and shellfish harvesting areas[2, 4]. Faecal indicator bacteria (FIB) are commonly used to determine water contamination of public health significance [5]. The identification and enumeration of indicator bacteria has several advantages. However, these methods fail to detect the source of faecal contamination[4]. Detection of the source is a pre-requisite for the effective and efficient management of these aquatic environments. It also reduces the time and cost of implementing remedial measures[6]. The intestines of warm-blooded animals contain abundant indicator bacteria, and their presence in environmental waters indicates faecal contamination, including the potential presence of pathogenic micro-organisms[7]. Van Asperen *et al.* [8] identified the quality of water which is necessary to protect bathers from illness and the key feature is the concentration of faecal contamination in the water. The EU Bathing Water Directives 2006/7/EC are mainly focused on the protection of the health of bathers and those involved in coastal recreation[5, 9-11]. In the UK normally the bathing season starts from 15 May and ends on 30 September [12, 13]. The objectives of this study were to monitor FIB (Enterococci and *E. coli*) and *Bacteroides spp.* in and out of the bathing season at a site of known problems noting the trends and to evaluate *Bacteroides* PCR-based tracking to source human faecal contamination. Sediment was also monitored as a possible reservoir of water contamination.

2 MATERIALS AND METHODS

2.1 The Study Area

South Sands, Salcombe is situated in the South Hams district of Devon and in the lower reaches of the Kingsbridge Estuary (latitude 50°13'N longitude 3°47'W). South Sands is a beautiful sandy beach and well known for human recreational activities which has experienced some FIB contamination in recent years although it generally meets the requirements for EU bathing waters. This study concentrates on three locations of the South Sands waters and sediments. These sites consist of the beach (A), the pond near to the hotel (B) and the stream near to the caravan park (C) (Figure 1).

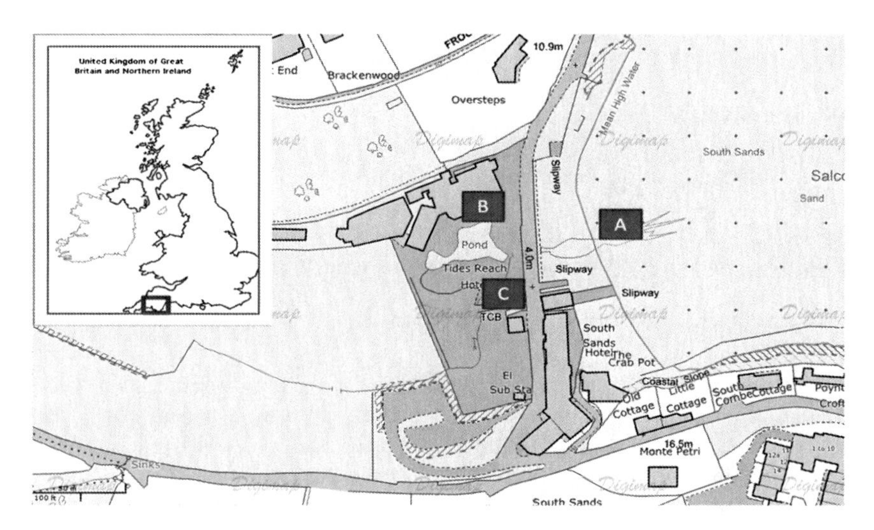

Figure 1 *The study area, South Sands, Salcombe, Devon. A: South Sands beach, B: Pond near the hotel and C: Stream just below the caravan park.*

2.2 Sampling

The water and sediment samples were taken (three samples from each site) in sterilized, non-toxic, wide mouth aseptic and labelled containers, at a depth of approximately 30 cm below the surface (for water), or from the sediment surface. The sampling containers were packed in an ice box, and these samples returned to the microbiological laboratory within six hours for further examination[14] and were analysed for the detection and enumeration of faecal indicator bacteria according to the EU Directive's Guidelines 2006/7/EC [11]. Human faeces were collected from healthy volunteers, whereas animal faeces (cow, horse, pig, sheep, deer and duck) were obtained from farms close to the study area. Water and sediment samples were obtained on four occasions in June, August, October 2010 and February 2011.

2.3 Filtration Method

The membrane filtration method was used to process the water and the sediment samples, and subsequent culture to detect and enumerate Enterococci, *E. coli* and *Bacteroides*. 100ml of the water was pulled through filter membranes (0.45 µm, Whatman, UK) with suction by a negative pressure vacuum pump. Two grams of sediment sample were taken and added to 20 ml sterile sea water (Sigma Fluka, UK). The sediment samples were blended (Seward Lab, UK) for 2 minutes, and left to settle for 10 minutes prior to aspirating the supernatant fluid. This was filtered through the membrane filters as previously [15, 16]. After filtration, the membranes were inoculated on Slanetz and Bartley agar to detect Enterococci (35 ^{0}C for 4 hrs for resuscitation, then 44 ^{0}C for 44 hrs), on *Bacteroides* Bile Esculin agar to detect *Bacteroides spp.* (37 °C for 48-72 hrs at anaerobic condition, Don Whitley, UK), and on Membrane Lauryl Sulphate Broth to detect *E. coli* (35 °C for 4 hours for resuscitation, then at 44 °C for 44 hours). Results were calculated in CFU/100ml for the water samples and CFU/g for the sediment samples[15, 17, 18].

2.4 DNA Extraction

DNA was extracted from the water and faeces samples by using QIAamp Stool DNA mini Kit (Qiagen, UK) following the manufacturer's protocol. The SoilMaster™ DNA Extraction Kit (Cambio, UK) was used for DNA extraction from the sediment samples following the manufacturer's protocol. DNA was stored at stored at -80 C° [19] before use.

2.5 PCR Technique

The Polymerase Chain Reaction (PCR) was used to detect *Bacteroides-Prevotella* 16S rDNA gene in the water, the sediment and faecal samples by using previously designed primer pairs (Table 1) [20, 21]. The PCR was carried out in a total volume of 25 µl reaction mixture containing 5 µl of template (extracted) DNA, 2 mmol l^{-1} $MgSO_4$, 0.2 mmol l^{-1} of each deoxynucleoside triphosphate, 0.4 µmmol l^{-1} of each primer (eurofins, MWG, Germany), 12.5 µl *Taq* DNA polymerase with buffer (Sigma Aldrich, UK). The cycling parameters were 15 minutes at 95 °C for initial denaturation, after that 35 cycles of 94 °C for 30s, annealing temperature as in the Table 1, 1.5 minutes at 72 °C, followed by extension at 72 °C for 7 minutes [20].

Table 1 *The host-specific Bacteroides 16S rDNA primers used in this study.*

Primer	Primer sequences	Annealing temp.(°C)	Length	Amplicon size (bp)	References
HF183F	ATCATGAGTTCACATGTCCG	55.3	20	525	Bernhard and Field [20]
Bac32F	AACGCTAGCTACAGGCTT	53.7	18	676	Bernhard and Field[20]
CF128F	CCAACYTTCCCGWTACTC*	54.8	18	580	Bernhard and Field [20]
HoF795F	CAAGCCGTAAAATAGTCG G	56.7	19	129	Dick *et al.* [22]
PF163F	GCGGATTAATACCGTATGA	52.4	19	563	Dick *et al.* [22]
Bac708R	CAATCGGAGTTCTTCGTG	With forward	18	With forward	Bernhard and Field [20]

* W: A or T, Y: C or T

2.6 Statistical Analysis

The results were statistically evaluated using SigmaPlot version 11.0 which included Kruskal-Wallis One Way analysis of variance on ranks (ANOVA). A *P*-value of 0.05 or less is considered significant.

3 RESULTS

3.1 Bacterial Culture

The bathing water at South Sands beach shows 'excellent' values for FIB (Enterococci and *E. coli*) at all times of sampling. However, water values of FIB and *Bacteroides* are significantly higher in the stream and higher still in the pond (Figure 2) at all times. All FIB and *Bacteroides* show a significant increase in values between June and August and between those samples collected in the bathing season and in the winter months. The highest water values are seen in October for Enterococci and in February for *E. coli* and *Bacteroides*. All sediment samples show a loading of both FIB and *Bacteroides* which also shows a significant increase out of the bathing season. The loading of pond to stream to beach is also reflected in the FIB/*Bacteroides* sediment values, the beach being the least contaminated.

3.2 Bacterial Source Tracking

Results of the PCR-based source tracking are shown in Table 2. The method was reproducible and results easy to interpret. The general *Bacteroides* primer set Bac32F and Bac708R were used to detect and confirm the *Bacteroides spp.* in the water, sediment and faecal samples: all samples gave a product at 670 bp (Figure 3a). The human specific primer HF183F only gave a positive report in the stream sediment at 520bp (Figure 3b). The cow primer CF128F gave a positive reaction with water and sediment of the stream and sediment of the beach at 580bp (Figure 3c). The horse primer (HoF597F) and pig primer (PF163F) gave negative results with all water and sediment samples (not shown). All primers gave positive results with their corresponding faecal sample with only the cow primer showing a lack of specificity by a positive reaction with the horse faecal sample (not shown).

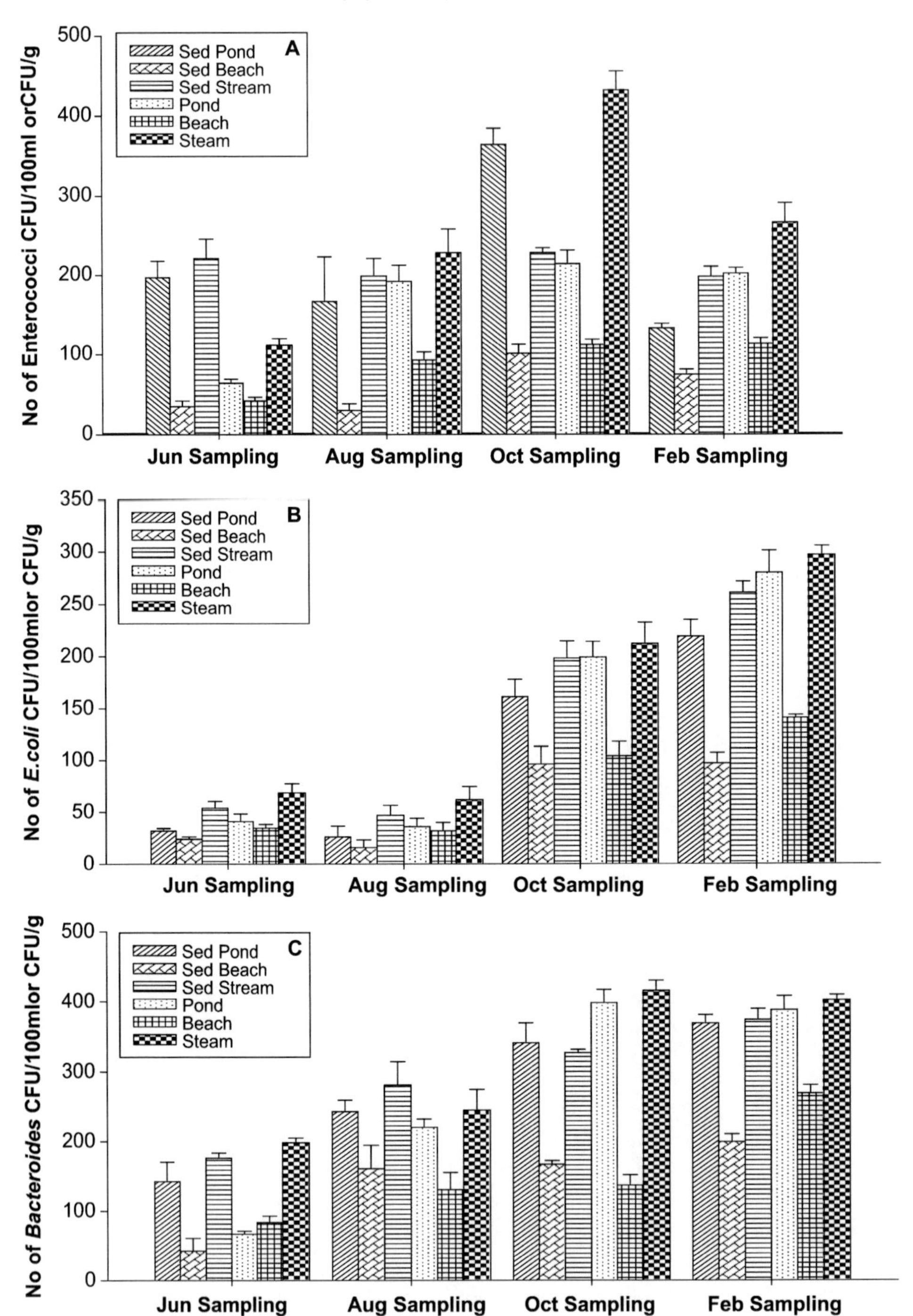

Figure 2 *The distribution of A: Enterococci, B: E. coli and C: Bacteroides spp. in South Sands water and sediment (mean and SD of triplicate samplings) collected in Jun, Aug, Oct and Feb.*

Table 2 *PCR host-specific Bacteroides primers in the water (W), sediment (S) and faecal samples.*

Primers	Stream		Pond		Beach		Human Faeces	Cow Faeces	Horse Faeces	Pig Faeces	Duck Faeces	Deer Faeces	Sheep Faeces
	W	S	W	S	W	S							
HF183F	-	+	-	-	-	-	+	-	-	-	-	-	-
Bac32F	+	+	+	+	+	+	+	+	+	+	+	+	+
CF128F	+	+	-	-	-	+	-	+	+	-	-	-	-
HoF597F	-	-	-	-	-	-	-	-	+	-	-	-	-
PF163F	-	-	-	-	-	-	-	-	-	+	-	-	-

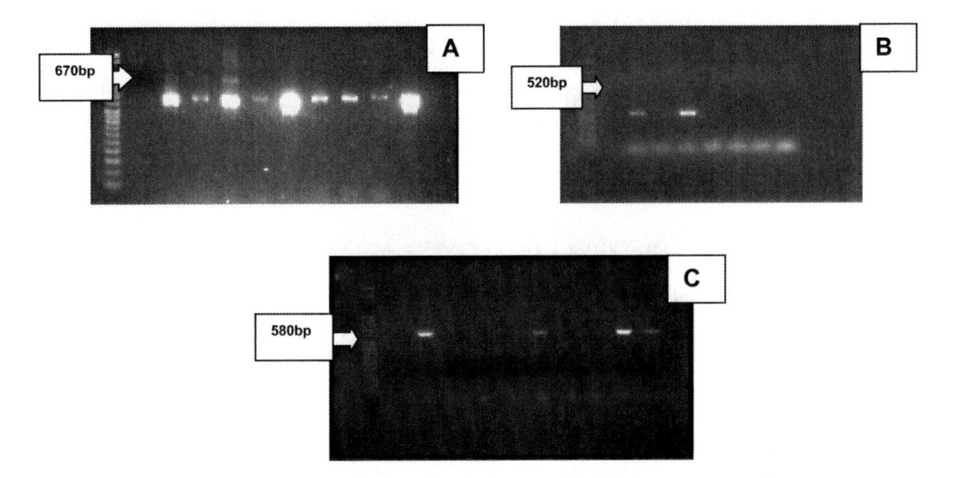

Figure 3 *Host-specific PCR amplified with general Bacteroides primer (Bac32F) (A), human specific primer (HF183F) (B) showing positive with the stream sediment, and cow primer (CF128F) (C) showing positive results with water and sediment from the stream and the sediment from the beach.*

4 DISCUSSION

4.1 Faecal Indicator Bacteria

The results of this study revealed that the quality of the bathing water was excellent or good based on the EU Bathing |Water Directive 2006 for Enterococci and *E. coli*. Higher bacterial contamination at the pond and the stream water and the sediment was shown in the results and a significantly higher concentration ($p<0.05$) than at South Sands beach. It is supposed that these sites (stream and pond) face more faecal contamination from different agriculture sources with animal faeces such as cow, sheep, deer, horses, ducks, and birds associated with this area. From the results, it appears that *Bacteroides spp.* and Enterococci isolates were always present in significantly higher concentrations ($p<0.05$) than *E. coli*. This is in agreement with Rees *et al.* [12] who found the concentration of FIB

(Enterococci and *E. coli*) on Spanish, Greek and Italian beaches were lower than on six UK beaches studied and that they were more variable. Also, most of those beaches met the EU Standards Directive 2006. Runoff from agricultural land and pasture land was identified as a significant source of these FIB[12, 15, 23, 24].

The sediment samples collected in this study also had significantly more bacterial contamination ($p< 0.05$) at the stream and pond than on South Sands beach which may be associated with the sediment type, as the beach is composed of mobile sand. These sediments could be a large potential source of faecal indicator bacteria as well as pathogenic microorganisms found in the aquatic environment[16, 25, 26]. Goyal *et al.* [27] and Ferguson *et al.* [16] refer to the re-suspension of sediments, which occurs due to runoff, animal traffic and storms, and can increase the concentration of the FIB in water. The number of bacteria in both water and sediment increases with heavy rainfall and the sediment acts as reservoirs for different types of the bacteria. The significant difference ($p< 0.05$) between the sampling occasions (June, August, October 2010 and February 2011) with October and February samplings having higher concentrations of the FIB than June and August samplings may be associated with the pattern of rainfall. Kay *et al.* [5, 28, 29] demonstrated that rainfall will facilitate the transport of sediment with attached bacteria and help to release the entrapped bacteria from the sediment. Kay *et al.* also reported that 40-50% of the faecal indicator bacteria were determined as agricultural sources such as livestock-voided faeces and spread manure, farmyard and manure heaps, and adjacent water [30].

4.2 PCR

The obligate anaerobic bacteria *Bacteroides* are preferred as host-specific markers because they are restricted to warm-blooded animals including humans and make up a large number of faecal bacteria: also they can only survive a short time in waters [31]. Bernhard and Field [21] demonstrated that the use of molecular methods such as PCR against culture-based methods enhanced the ability for their use in water quality detection. In our study all the water and the sediment samples appeared positive with the general *Bacteroides* markers indicating some faecal pollution. The requirement of *Bacteroides spp.* to live under anaerobic conditions suggests that all sampling sites were regularly subjected to contamination recently by the faecal materials [32]. However, the persistence of the PCR markers in the environment is not known to any great degree and needs further study. The results displayed human origin contamination in the sediment of the stream indicating that human faecal contamination may be occurring upstream, possibly from a storm sewer overflow. Gawler *et al.* [33] have demonstrated that this HF183 marker shows a high sensitivity (100%) in all four Atlantic Rim countries of the European Union (France, Ireland, Portugal and UK), in the USA (100%) [20], in Canada (94-100%) [32] and in Australia[19, 34].

The cow marker (CF128F) showed a positive finding in the stream water, the stream sediment and the beach sediment indicating that the majority of the FIB contamination here is by cow faeces from upstream. Caution should be taken here, however, since this marker was also present in the horse faeces tested here. Gawler *et al.* [33] reported that the CF128F marker displayed a high sensitivity (100%) but is a marker for ruminant faeces rather than bovine specific. In addition, the CF128F marker is present in pig faeces in France, Portugal and UK. It is also in human and chicken faeces in Portugal, which suggests a limitation this marker [33]. On the other hand, no horse and pig markers appeared in any water or sediment sample, indicating no faecal contamination from either horse or pig. In addition, Dick *et al.* [22] demonstrated that the horse and pig markers showed a high sensitivity (90%, 100%) with horse and pig faeces respectively.

5 CONCLUSIONS

The culture-based results of this study revealed that whilst FIB (Enterococci and *E. coli*) and *Bacteroides spp.* were present in all water and sediment samples from South Sands, the bathing water met the EU Directive 2006 standards. However the water quality of all samples deteriorated out of the bathing season. Sediment samples, especially those of the stream and pond, could be a reservoir for 'spikes' which occur in the FIB under certain conditions. The *Bacteroides* 16S rDNA host-specific PCR was successful in showing that although human faecal markers were not present on the beach they were present in the stream indicating a possible hazard, the positive cow marker CF128F in water and sediment of the stream indicates that FIB are primarily due to cow faecal contamination at this site. The persistence of the *Bacteroides* markers in the environment is not known to any great degree and needs further study along with confirmation of their specificity.

Acknowledgements

We would like to thank the Ministry of Higher Education and Scientific Research of Iraq for funding of this study.

References

1. A. J. A. Vinten, G. Sym, K. Avdic, C. Crawford, A. Duncan and D. W. Merrilees, *Water Research*, 2008, **42**, 997.
2. B. Ray, *Injured index and pathogenic bacteria: occurrence and detection in food, water and feeds.*, CRC Press Inc, Florida, 1989.
3. D. Kay, A. Wither and A. Jenkins, *Faecal Pathogens*, Water Environ. Manag., Suffolk, 2000.
4. T. M. Scott, J. B. Rose, T. M. Jenkins, S. R. Farrah and J. Lukasik, *Appl. Environ. Microbiol.*, 2002, **68**, 5796.
5. D. Kay, J. Crowther, L. Fewtrell, C. A. Francis, M. Hopkins, C. Kay, A. T. McDonald, C. M. Stapleton, J. Watkins, J. Wilkinson and M. D. Wyer, *Environ. Sc. Policy*, 2008, **11**, 171.
6. S. Okabe, N. Okayama, O. Savichtcheva and T. Ito, *Appl. Microbiol. Biotechnol.*, 2007, **74**, 890.
7. K. L. Anderson, J. E. Whitlock and V. J. Harwood, *Appl. Environ. Microbiol.*, 2005, **71**, 3041.
8. I. A. Van Asperen, G. Medema, M. W. Borgdorff, M. J. W. Sprenger and A. H. Havelaar, *Internation. J. Epidemiol.*, 1998, **27**, 309.
9. D. Kay, M. Aitken, J. Crowther, I. Dickson, A. C. Edwards, C. Francis, M. Hopkins, W. Jeffrey, C. Kay, A. T. McDonald, D. McDonald, C. M. Stapleton, J. Watkins, J. Wilkinson and M. D. Wyer, *Environ. Pollution*, 2007, **147**, 138.
10. P. Kay, A. C. Edwards and M. Foulger, *Agricul. Syst.*, 2009, **99**, 67.
11. EU, *Offic. J. Eu. Un.*, 2006, **L64**, 37.

12. G. Rees, K. Pond, K. Johal, S. Pedley and A. Rickards, *Water Research*, 1998, **32**, 2335.
13. W. Howarth and D. McGillivary, *Water Pollution Water Qual. Law*, Shaw and Sons, Kent, 2001.
14. G. Bradley, J. Carter, D. Gaudie and C. King, *Journal of Applied Microbiology*, 1999, **85**, 90.
15. D. L. Craig, H. J. Fallowfield and N. J. Cromar, *J. Appl. Microbiol.*, 2002, **93**, 557.
16. D. M. Ferguson, D. F. Moore, M. A. Getrich and M. H. Zhowandai, *J. Appl. Microbiol.,* 2005, **99**, 598.
17. E. J. Fricker and C. R. Fricker, *J. Microbio. Meth.*, 1996, **27**, 207.
18. S. J. Livingston, S. D. Kominos and R. B. Yee, *J. Clin. Microbiol.*, 1978, **7**, 448.
19. W. Ahmed, J. Stewart, D. Powell and T. Gardner, *Letters Appl. Microbiol.*, 2008, **46**, 237.
20. A. E. Bernhard and K. G. Field, *Appl. Environ. Microbiol.*, 2000, **66**, 4571.
21. A. E. Bernhard and K. G. Field, *Appl. Environ. Microbiol.*, 2000, **66**, 1587.
22. L. K. Dick, A. E. Bernhard, T. J. Brodeur, J. W. Santo Domingo, J. M. Simpson, S. P. Walters and K. G. Field, *Appl. Environ. Microbiol.*, 2005, **71**, 3184.
23. Y.-J. An, D. H. Kampbell and G. Peter Breidenbach, *Environ. Pollution*, 2002, **120**, 771.
24. J. Crowther, D. Kay and M. D. Wyer, *Water Research*, 2002, **36**, 1725.
25. J. B. Ellis, *Chemosphere*, 2000, **41**, 85.
26. L. P. Joseph, Univ. Plymouth, 2009.
27. S. M. Goyal, C. P. Gerba and J. L. Melnick, *Appl. Environ. Microbiol.*, 1977, **34**, 139.
28. D. Kay, M. D. Wyer, J. Crowther, J. Wilkinson, C. Stapleton and P. Glass, *Water Research*, 2005, **39**, 3320.
29. D. Kay, A. Edwards, R. Ferrier, C. Francis, C. Kay, L. Rushby, J. Watkins, A. McDonald, M. Wyer and J. Crowther, *Prog. Physic.Geograph.*, 2007, **31**, 59.
30. D. Kay, S. Anthony, J. Crowther, B. J. Chambers, F. A. Nicholson, D. Chadwick, C. M. Stapleton and M. D. Wyer, *Sc. Total Environ.*, 2010, **408**, 5649.
31. C. L. Meays, K. Broersma, R. Nordin and A. Mazumder, *J. Environ. Manag.*, 2004, **73**, 71.
32. B. Fremaux, J. Gritzfeld, T. Boa and C. K. Yost, *Water Research*, 2009, **43**, 4838.
33. A. H. Gawler, J. E. Beecher, J. Brandão, N. M. Carroll, L. Falcão, M. Gourmelon, B. Masterson, B. Nunes, J. Porter, A. Rincé, R. Rodrigues, M. Thorp, J. Martin Walters and W. G. Meijer, *Water Research*, 2007, **41**, 3780.
34. W. Ahmed, A. Goonetilleke, D. Powell and T. Gardner, *Water Research*, 2009, **43**, 4872.

DETECTION AND QUANTIFICATION OF *E. COLI* AND COLIFORM BACTERIA IN WATER SAMPLES WITH A NEW METHOD BASED ON FLUORESCENCE *IN SITU* HYBRIDISATION

Michael Hügler,[1] Karin Böckle,[1] Ingrid Eberhagen,[1] Karin Thelen,[2] Claudia Beimfohr,[2] and Beate Hambsch[1]

[1]DVGW-Technologiezentrum Wasser (TZW), Karlsruher Str. 84, D-76139 Karlsruhe, Germany. E-mail: michael.huegler@tzw.de, beate.hambsch@tzw.de
[2]vermicon AG, Emmy-Noether-Str. 2, D-80992 Munich, Germany

1 INTRODUCTION

Safe drinking water must be free of pathogenic microorganisms. Since the presence of pathogens correlates well with the presence of faecal contaminations, testing for microbial indicator organisms is an option used to monitor microbiological contaminants in water supplies.[1] Up to date the monitoring of *E. coli* bacteria as a specific indictor for faecal contamination is of capital importance in drinking water as well as in recreational waters (e.g. bathing water, beaches, springs etc.).[2] In addition, coliform bacteria are frequently used as an indicator of microbiological water quality.[3,4] Coliform bacteria are not necessarily of faecal origin, as some genera are also found in natural environments.[4] Yet, their presence in drinking water must be considered as a possible threat or at least to be indicative of microbiological water quality deterioration, since drinking water is not a natural growth environment for coliform bacteria.[3] Other microbial indicators of importance are e.g. Enterococci, Clostridia or bacteriophages and many alternative indicators are discussed in the literature.[5,6]

Regardless of which indicator organism or pathogen is monitored, fast and sensitive methods are required.[7] The standard cultivation methods are often too time-consuming to match the requirements of modern water safety management, i.e. approaching the equivalent of online measurements. The Fluorescence *in situ* Hybridization (FISH) technology is a fast molecular method, which can be used for the cultivation-independent, rapid, sensitive and specific detection of single microbial cells.[8-10]

Within the TECHNEAU project – an integrated project funded by the European Commission – a protocol for the simultaneous detection of *E. coli* and coliform bacteria named "ScanVIT-E.coli/coliforms" was developed.[11] The protocol consists of two different approaches. One approach allows the direct detection of single *E. coli* and coliform bacterial cells on filter membranes. The second approach includes an incubation step on a nutrient agar plate and subsequently the detection of the grown micro-colonies. Both approaches were validated using drinking water samples spiked with pure cultures and naturally contaminated water samples. Furthermore, the effects of heat, chlorine and UV disinfection on the FISH based detection of *E. coli* and coliform bacteria have been investigated.[11]

Within this chapter, we briefly describe the developed method; we summarize its pros and cons, and present the results of the validation studies.[11] We furthermore discuss a method for the automated quantification of the fluorescent micro-colonies.

2 METHOD AND RESULTS

2.1 Development of the FISH protocol

Precise quantification is a requirement for any method used for the analyses of *E. coli* and coliform bacteria in drinking water and is the assumption inherent in limit values defined in national and international guidelines and regulations. Therefore, within the developed kit (ScanVIT-E.coli/coliforms; vermicon AG, Munich, Germany), the FISH staining is applied to filter membranes after filtration of a defined volume of water. Two different approaches are possible: either the direct analysis of single cells on the filter membrane (approach 1) or an approach combining an incubation step (6-7 h at 37°C) on a nutrient agar plate (lactose-TTC-agar) prior to FISH staining of the grown micro-colonies (approach 2). Two probes labelled with different fluorescent dyes are implemented in the kit, a probe with a red-fluorescent dye for coliform bacteria, and a probe with a green-fluorescent dye for *E. coli*.[11] Figure 1 shows the application of both approaches to a drinking water sample spiked with a pure culture of *E. coli* cells.

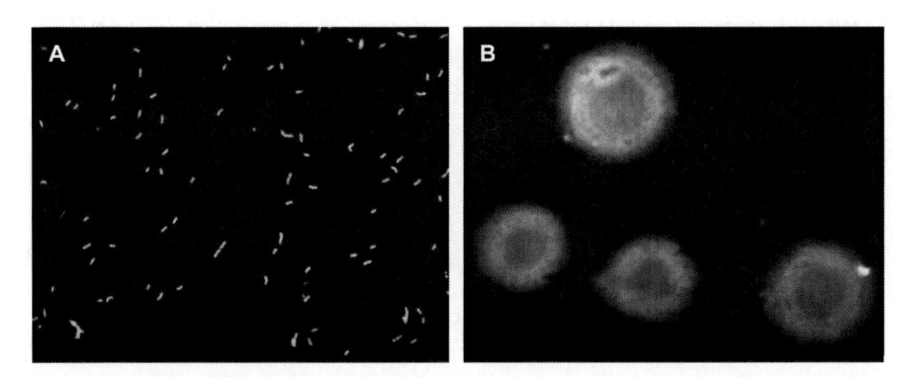

Figure 1 *Microscopic pictures of E. coli cells and micro-colonies labelled with the green fluorescent E. coli probe. (A) Single E. coli cells (size approx. 1 µm), (B) micro-colonies (size > 100 µm).*

2.2 Validation of the single cell protocol

For the validation of the single cell protocol, water was spiked with pure cultures of *E. coli* or *Citrobacter freundii* cells. The effects of heat and disinfection treatments (chlorine, UV) were investigated by comparing fluorescent cell counts obtained by the FISH method to total cell counts (acridine orange staining) and colony counts after cultivation on chromocult coliform agar (for details see Hügler et al. 2011).[11]

In principle, the single cell FISH-method worked well. In samples which have not been disinfected, fluorescent cell counts determined after FISH staining were in the same range as colony counts after cultivation on chromocult coliform agar. The same holds true for heat treated samples.[11] However, the single cell approach showed limitations when disinfected samples were studied.

When UV irradiation was used as disinfection method, UV treated *E. coli* or *C. freundii* cells could still be hybridized with the respective FISH probes, although they could not be cultured anymore.[11] Thus, the rRNA – target of the FISH probe – still seems to be intact after UV treatment, while cultivability is impaired. This leads to a discrepancy in results between both methods. At this time, it is not clear to what extend the cells have been damaged by UV irradiation. There are indications, that the cells were permanently damaged, as we failed to resuscitate and re-grow the UV treated cells. Further analyses are needed to clarify this issue.

The effects of chlorine disinfection were also investigated with *E. coli* and *C. freundii* cells. As expected, after treatment with chlorine solution (6 mg/L, 30 min), the cells could not be cultured on chromocult coliform agar anymore. Yet, the cell number did not decrease, as shown by analysis of the total cell counts after acridine orange staining, indicating that the cellular DNA was still somehow intact after chlorination. In contrast, the experiments indicated decomposition of the rRNA molecules by the impact of chlorine, as the cells could not be detected in the fluorescent microscope after application of the red-labelled coliform probe.[11] However, all cells (FISH stained and not stained, target as well as non-target cells) emitted an unspecific green fluorescence after chlorine treatment at the excision wavelength of the green fluorescent dyes. Due to this effect, false-positive results can be obtained when green fluorescent dyes are used with chlorine treated samples.

Generally speaking, the single cell approach is suitable for the analyses of non-disinfected water samples with a broad range of concentrations of *E. coli* and coliform bacterial cells. However, one has to keep in mind, that for small cell numbers, microscopic evaluation can become very time- and labour-intensive as, due to the low concentrations in drinking water, the whole filter membrane must be scanned in order to detect every single fluorescent cell (for instance 1 per 100 mL in drinking water). Automated quantification would be advantageous to make this approach practicable. Within the project, it was not possible to develop automated quantification with laser based or microscope based systems. Furthermore, target cells can hardly be detected if a huge amount of background flora is present (e.g. in surface water samples with a cyanobacterial bloom).

2.3 Validation of the micro-colonies protocol

Approach 2 of the FISH protocol for the detection of *E. coli* and coliform bacteria in water samples includes an incubation step on a lactose-TTC-agar plate for 6 to 7 hours in advance of the application of the FISH probes. Thus, in this approach, the grown micro-colonies are stained rather than single bacterial cells (Figure 1).

For the validation study, drinking water samples and natural water samples (e.g. river water, waste water) were analyzed.[11] The upper and lower detection limit of the approach was tested by spiking drinking water samples with varying numbers of *E. coli*, *C. freundii*, *Enterobacter cloacae* and/or *Serratia marcescens* cells. The used concentration range was between 1 and 10^6 cells/100 mL. Colony counts of the FISH-stained micro-colonies were compared to colony counts after cultivation on chromocult coliform agar. Both methods showed congruent numbers in the low as well as in the high cell concentration range (up to approx. 10^5 cells/100 mL) (Figure 2).[11]

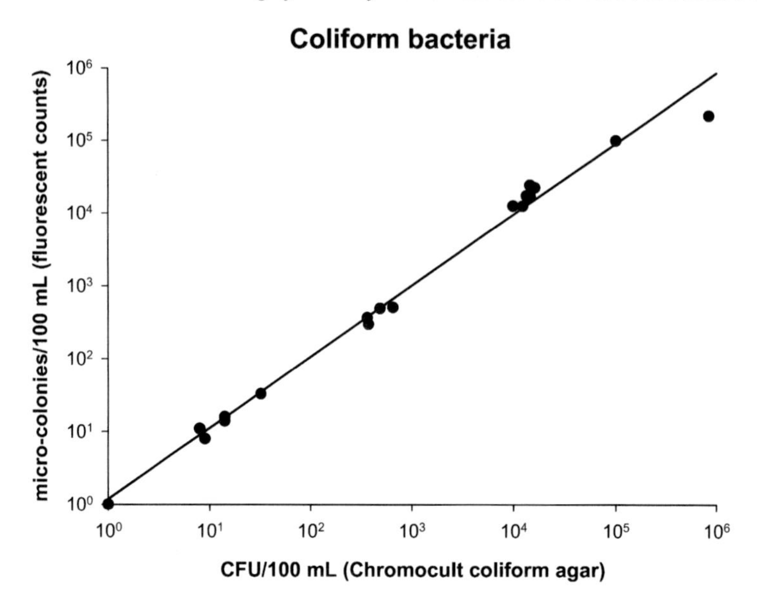

Figure 2 *Correlation between micro-colony counts after FISH and colony counts on chromocult coliform agar for coliform bacteria. The solid line indicates the 1:1 relationship (adapted from Hügler et al. 2011).*[11]

The effects of disinfection methods were investigated by treating the spiked water samples with chlorine (0.5 or 6 mg/L, for 30 min) or UV light (450 J/m^2, for 15 min). As expected, no growth could be observed in the disinfected samples, neither with the micro-colony FISH approach, nor with the direct culturing method.[11]

The validation study included also the investigation of natural water samples harbouring elevated concentrations of *E. coli* and coliform bacteria. Some of the results are shown in Table 1. The effect of the water matrix (i.e. the organic carbon present in natural waters) was studied by either direct filtration of the natural water sample with its matrix, or by adding the enriched mixed natural biocoenosis to drinking water (see Hügler et al. 2011 for methodological details).[11] As shown in Table 1, we could not detect any differences between these samples, indicating that the water matrix did not interfere with the micro-colony approach.

Furthermore, the micro-colony approach was successfully used to study the concentration of *E. coli* and coliform bacteria in water samples from the different treatment steps of the New Goreangab Water Reclamation Plant (Windhoek, Namibia). In this plant, drinking water is directly reclaimed from treated waste water effluent by a serious of state-of-the-art treatment steps.[12] The FISH method worked well for all of the samples retrieved, even when large numbers of non-target microorganisms (background flora) were present.

Table 1 *Comparison of colony counts and fluorescent micro-colony counts after FISH in natural water sample.*

Water sample and bacteria used for spiking	*colony counts on chromocult coliform agar CFU/100 mL coliform bacteria*	*(E. coli)*	*micro-colony counts after FISH micro-colonies/100 mL coliform bacteria*	*(E. coli)*
river water (Rhine)	364	(11)	366	(11)
river water (Rhine) + *E. coli*	374	(24)	300	(20)
river water (Main)	16100	(1200)	22400	(1390)
waste water	840000	(5700)	220000	(5700)
drinking water + mixed biocoenosis[a]	488	(11)	492	(11)
drinking water + mixed biocoenosis[a] + *E. coli*	652	(18)	508	(25)

[a] from river water (Rhine)

2.4 Automated detection and quantification of micro-colonies with microscope based systems

In order to make the FISH method practicable for routine analyses of *E. coli* and coliform bacteria in water samples, automated identification and quantification of the micro-colonies on the filter membrane would be required. To facilitate automated quantification, the entire filter membrane should be displayed in one image (Figure 3, A). For the image generation, a Leica MS5 fluorescence-stereomicroscope (Leica, Wetzlar, Germany) equipped with an achromat-objective, a dual-band fluorescence filter (for parallel excitation of red and green fluorescent dyes) and a digital video-camera for documentation was used. A macro-based software application (QWIN V3 Runner, Leica) was used for the subsequent quantification of the micro-colonies.

After FISH hybridization, an image of the complete filter membrane was taken and subsequently analysed by the developed software macro. The software macro consisted of six successive steps (shown in Figure 3). After image selection (Figure 3, A), the evaluation area was selected (Figure 3, B, the circle corresponds to the evaluation area). Subsequently, the software-macro transferred the RBG image into a grey-scale image for manual correction of brightness and contrast (Figure 3, C, signal-to-noise optimization). For better visualization, the area of the micro-colonies was coloured by the software macro (Figure 3, D). In an optional step, adherent micro-colonies could now be separated manually (Figure 3, E). Finally, the coloured micro-colonies were automatically quantified by the software (Figure 3, F).

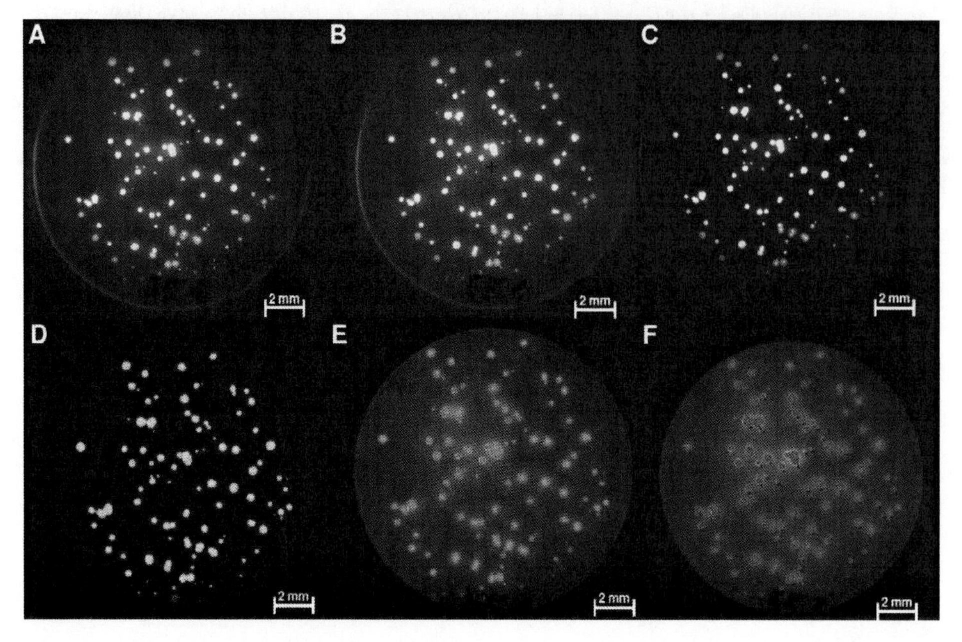

Figure 3 *Main steps of the software macro for the quantification of micro-colonies on filter membranes (see text for details).*

The reliability of the developed software-macro was evaluated by comparison of the results using semi-automated quantification to manual quantification (Table 2). The deviation between the two quantification methods was in a range of +/- 5 % (absolute deviation maximum: 6 colonies). Thus, indicating that the software macro based on semi-automated quantification of micro-colonies is as reliable as manual quantification of micro-colonies.

Table 2 *Comparison of semi-automated and manual quantification of micro-colonies on filter membranes.*

Sample No.	*automated quantification (colonies/filter)*	*manual quantification (colonies/filter)*
1	302	303
2	121	120
3	25	25
4	67	68
5	81	81
6	0	0
7	6	6
8	44	42
9	263	267
10	171	165

3 CONCLUSION

The standard slide-based FISH protocol has been successfully transferred to a filter membrane-based protocol and the simultaneous and specific detection of *E. coli* and coliform bacteria has been established. Two different detection and quantification strategies – single cell and micro-colony detection – were implemented.

The single cell FISH-method is very fast, it takes approx. 2 h from sample filtration to microscopic evaluation. However, microscopic evaluation can become very labour-intensive as the whole membrane filter must be scanned manually with high magnification. This holds especially true for low cell numbers. Automated cell detection and counting is necessary to make this approach useful for routine analyses. So far, such automation has not been successfully implemented for the single cell approach. Furthermore, the approach showed some drawbacks when samples, which had been disinfected by UV irradiation or chlorine, were analyzed.[11]

The micro-colony approach yielded very good results for all different samples and conditions tested and, thus, can be thoroughly recommended for usage as an alternative method to detect *E. coli* and coliform bacteria in water samples. The method is not restricted to drinking water samples, but can also be used in all kinds of water samples (e.g. recreational waters, waste water, and surface water). The approach has the potential for practical routine use, as manual quantification is feasible and automated quantification can be implemented. As micro-colonies have a bigger size than single cells, only a low magnification factor is needed for enumeration, making even manual quantification not too time-consuming. Due to the incubation step of 6-7 h, it takes approximately 10 h to get results with this method, making it more time-consuming than the single cell approach.

This demonstrates that the novel method developed for the detection of *E. coli* and coliform bacteria with the FISH technique is a promising and robust technology, which can be applied for routine analyses of water samples. The time requirement for the entire procedure (filtration, incubation, FISH labelling and automated evaluation) is less than one working day, even if the more time-consuming micro-colony approach is used. Hence the method is considerably faster than traditional solely culture-based methods.[13-16]

Acknowledgements

The authors acknowledge financial support by the 6th EU framework project TECHNEAU (Technology Enabled Universal Access to Safe Water, contract number 018320).

References

1 M. Stevens, N. Ashbolt and D. Cunliffe, *Review of Coliforms as Microbial Indicators of Drinking Water Quality*, National Health and Medical Research Council, Canberra, Australia, 2003.
2 P. Tallon, B. Magajna, C. Lofranco and K.T. Leung, *Water Air Soil Pollut.*, 2005, **166**, 139.
3 A. Rompré, P. Servais, J. Baudart, M.E. de-Robin and P. Laurent, *J. Microbiol. Meth.*, 2002, **49**, 31.
4 H. Leclerc, D.A.A. Mossel, S.C. Edberg and C.B. Struijk, *Annu. Rev. Microbiol.*, 2001, **55**, 201.

5 J.P.S. Cabral, *Int. J. Environ. Res. Publ. Health*, 2010, **7**, 3657.
6 M.J. Figueras and J.J. Borrego, *Int. J. Environ. Res. Publ. Health*, 2010, **7**, 4179.
7 F. Sacher and B. Hambsch, `State-of-the-art in drinking water monitoring′ in *TECHNEAU: Safe drinking water from source to tap*, eds. T. van den Hoven and C. Kazner, IWA Publishing, London, 2009, pp.135-143.
8 R.I. Amann, I. Krumholz and D.A. Stahl, *J. Bacteriol.*, 1990, **172**, 762.
9 R. Amann, W. Ludwig and K.H. Schleifer, *Microbiol. Rev.*, 1995, **59**, 143.
10 R. Amann and B.M. Fuchs, *Nat. Rev. Microbiol.*, 2008, **6**, 339.
11 M. Hügler, K. Böckle, I. Eberhagen, K. Thelen, C. Beimfohr and B. Hambsch, *Wat. Sci. Technol.*, 2011, **64**, 1435.
12 J. Menge, `Treatment of wastewater for re-use in the drinking water system of Windhoek′ in *Proceedings of the Water Institute of Southern Africa (WISA)*, 2006.
13 J.K. Cowburn, T. Godall, E.J. Fricker, K.S. Walter and C.R. Fricker, *Let. Appl. Microbiol.*, 1994, **19**, 50.
14 E.J. Fricker, K.S. Illingwort and C.R. Fricker, *Wat. Res.*, 1997, **31**, 2495.
15 F.M. Schets, P.J. Nobel, S. Strating, K.A. Mooijman, G.B. Engels and A. Brouwer, *Let. Appl. Microbiol.*, 2002, **34**, 227.
16 C. Bernasconi, G. Volponi and L. Bonadonna, *Wat. Sci. Technol.*, 2006, **54**, 141.

A REVIEW OF POTENTIAL CULTURE INDEPENDENT BIOLOGICAL DETECTION METHODS FOR THE WATER INDUSTRY - CHALLENGES OF MOVING BEYOND THE RESEARCH LAB

Q.I. Sheikh[1], J.B. Boxall[2] and C.A. Biggs[3*]

[1]Department of Molecular Biology and Biotechnology, The University of Sheffield, Western Bank, Firth Court, Sheffield, S10 2TN, UK.
[2]Pennine Water Group, Department of Civil and Structural Engineering, The University of Sheffield, UK.
[3]Department of Chemical and Biological Engineering, The University of Sheffield, Mappin Street, Sheffield, S1 3JD, UK.
* Corresponding author Email: c.biggs@sheffield.ac.uk

1 INTRODUCTION

Contamination of drinking water with pathogens poses a serious threat to millions of people in the world. Even in today's developed world periodic outbreaks of diseases caused by polluted water is a severe public health problem. Our awareness regarding hygiene measures in drinking water has also increased dramatically over the last few decades. Thus, surveillance of drinking water is imperative to minimize such contaminations and ensure continuous supplies of healthy water world-wide.[1]

The hygienic quality and potential health risk from the consumption of drinking water is assessed by the cultivation of indicator bacteria as outlined by the European Union Council Directive 98/83/EC.[2] Current detection techniques used by the water industry are generally culture-based methods for indicators of faecal contamination. Although cultivation and quantification is carried out routinely using culture based indicator techniques, these methods are prolonged, involve multiple steps and, as a consequence of being an indicator technique, lack specificity.

In the revised draft of the drinking water directive (DWD) there is a general agreement that the current reference method is not suitable for a number of drinking water samples.[3] The need to change the current reference method is also recognised in The Water Safety Plan approach by the World Health Organisation (WHO)[4]. It has also been proposed that more than one method should be used.[5] The potential presence and quantity of microorganisms with high virulence in drinking water, therefore, requires changes in detection techniques beyond indicator methods to allow improvements in specificity and sensitivity. Investigations are also needed for the examination of microbial communities that may be vital from a human health standpoint.

In recent years, progress has been made in the field of pathogen detection by improving culture based methods. For example the use of selective enrichment broth (SVV) for simultaneous growth of *Salmonella*, *Vibrio parahaemolyticus*, and *Vibrio cholerae* by a single factor experiment and response surface method.[6] However, there still is a need for faster, more sensitive and specific detection methods.

The use of molecular microbiological detection and identification methods for pathogens in drinking water offers a promising alternative for the water industry and the US Environmental Protection Agency in 2008 published a review of molecular methods for detection of pathogens in water.[7] An accurate assessment of current techniques however is a prerequisite before comparison with new methods is feasible.

The objective of this review is, therefore, to describe developments in selected molecular techniques that show promise and present a smarter way for water borne pathogen detection over the current indicator methods. It will explain how some techniques are adapted for characterizing pathogenic microbial community composition and structure in variety of samples, but with a special focus on drinking water. We have structured this chapter by dividing the large variety of detection techniques into three main groups: whole cell based detection, biosensors and molecular detection methods. Newly developed techniques increased our insight into the vast diversity and interaction of microorganisms present in complex environments and, hence, we have given more attention on these advanced molecular microbiological methods and tried to encompass them.

2 WHOLE CELL BASED DETECTION METHODS

2.1 Microscopy and Staining Based Detection

Staining is a technique that is used to enhance microscopic images in microscopy. Pathogens in environmental samples can be detected using the addition of various stains such 4',6-diamidino-2-phenylindole (DAPI) and Hoechst stains, e.g. Hoechst 33258 and Hoechst 33342. These are part of a family of fluorescent stains for labeling deoxyribonucleic acid (DNA) in fluorescence microscopy and fluorescence-activated cell sorting (FACS). These stains can pass through cell membrane and bind strongly with DNA. Using an epifluorescent microscope it is then possible to count the number of cells that have been stained. Whilst such a technique provides quantitative analysis of the bacteria within the water samples, DAPI stains all DNA, pathogenic or not and therefore lacks specificity.

Increased specificity can be achieved with staining techniques such as fluorescent *in situ* hybridization (FISH) which have been used to detect pathogens such as *Streptococcus pyogenes*,[8] and oocysts.[9] FISH is an influential tool for assessing the expression and localization properties of DNA and ribonucleic acid (RNA) in whole organism and tissues. It could be used to provide both qualitative and quantitative information about specific pathogens within a water sample. Despite the promise offered by the combination of FISH and immunofluorescence (IF), in terms of reduced bench time and easy visualization of multiple labels at once, some technical hurdles have to be overcome to produce reliable data from these state-of-the-art neuroanatomy techniques.[10]

2.2 Non-Microscopy Based Detection

Flow cytometry is a technique for counting and examining microscopic particles, such as microorganisms, by suspending them in a stream of fluid and passing them by a detection apparatus. To assist in detection of bacteria and viruses, fluorescent dyes that attach to components of cells (such as nucleic acid) are added to the samples before they are analyzed using specialist laboratory equipment such as commercially available flow cytometers. Various modifications in this method have also been adapted such as flow cytometry combined with specific fluorescent antibodies as a rapid detection method for microorganisms for probing of cell surface antigens of target organisms.[11].

Flow cytometry has already been shown to be a valid tool for the rapid detection of the viability of pathogenic bacteria in food and environmental matrices to control and prevent health risks.[2] Flow cytometry was successfully used to detect the viability of *Aeromonas hydrophila* in waters and foods.[12] Keserue et al[13] also recently used flow cytometry in combination with immunomagnetic separation for detection of *Giarida lamblia* cysts. In a comparative study for the detection of the human pathogen *Pneumocystis jirovecii* using flow cytometry and routine techniques, flow cytometry was found to have a higher level of sensitivity.[14]

However, when tested, this technology seems unsuitable for enterococci detection in recreational waters [12] and further improvements in research and instrument design are needed to make this a viable technology suitable for pathogen detection across all types of water samples.[11] A recent example is the development of mass cytometry that overcomes the limitations of false readout by fluorescence spectral overlap in traditional fluorescence flow cytometers.[15] This new revolutionary flow cytometry based on mass detection has not directly been tested with pathogens in water samples, however the new development is comparable with the traditional flow cytometric methods and has potential for use in water industry.

Table I compares the culture based technique that is currently used for pathogen detection in the water industry with the whole cell detection based methods reviewed. Head to head comparison for key factors such as precision, cost per sample and time required provide an overview for the current limitation of their use.

Table 1. *Comparison of culture* vs. *culture independent whole cell detection methods.*

	Culture based	DAPI	FISH	Flow cytometry
Time	2 – 5 Days	1 Day	1- 2 Days	Hours
Cost / sample	Low capital costs (~£1 -£2 per sample)	Large capital costs (fluorescent microscopes) and consumables	Large capital costs (fluorescent microscopes) and consumables	Large capital costs (£10k to £100ks) for specialized equipment and consumables
Precision	Low	High (few cells/ml)	High (few cells/ml)	High
Specificity	Low	Low to Medium	Medium to High	Medium to High
Level of Expertise	Low	Medium	High	High
Type of information and sample	Grab samples Numbers – indicator organisms	Grab samples Numbers – all DNA	Grab samples Numbers – general or specific	Grab or potential for continuous Numbers – general or specific

3. BIOSENSORS

A relatively recent publication edited by Zourob et al[16] classified and described various emerging and novel biosensor technologies for rapid and sensitive detection of pathogens of concern to the food supply.[16] Biosensors-based methods however have the capability to meet the need for the detection of microbes and pathogens in different environmental samples, including drinking water.

A biosensor incorporates a biological component coupled to a transducer, which translates the interaction between the analyte and the biological component into a signal that can be processed and reported.[17] A wide range of transducers has been employed in biosensors, the most common of which are electrochemical and optical.[17] Biosensors therefore have the capacity to provide continuous measurements, ideally without the addition of reagents. Biosensors also have the potential of being highly useful in reducing the time for diagnosis and can have excellent sensitivity and specificity performance, allowing for the detection of contaminants.[18] Furthermore, they can be multiplexed into tens or hundreds of arrays, making the device a critical tool for rapid screening. The biosensor chips can also be miniaturized and are compatible with automated electronic data processing technologies, thus they can be integrated into small portable field-based devices and can be used off site.[18] Also, as manufacturing costs become cheaper the biosensors become more and more affordable.

In this section, we summarize potential biosensor based detection methods that have been successfully employed. The demand for biosensors to detect pathogens and pollutants is however growing day by day.[19] Biosensors will dominate the market in a very short time and could become an indispensable detection tool in the field and laboratory.[18] The use of biosensors as emerging technologies could therefore revolutionize the study and detection of pathogens. It is therefore hoped that biosensor techniques will play an extensive role in understanding the occurrence of contamination at the source during the next decade and help forecast the potential for risk and mitigation of outbreaks to occur.[18]

3.1 Bacteriophage-based Detection of Bacterial Pathogens

The use of bacteriophage or parts thereof for bacterial detection is attracting increasing attention, as reflected by a multitude of different phage-based techniques recently reported.[20] Bacteriophages are considered the most abundant organism and by far the most abundant self-replicating units on Earth.[20 21] They mediate interactions with bacterial hosts and, for decades, this has been exploited in phage typing schemes for signature identification of clinical, food-borne, and water-borne pathogens.[21] Phages or phage-derived proteins can therefore be used to detect the presence of unwanted pathogens in varying environments, which allows quick and specific identification of viable cells. Hence use of phages for detection of pathogenic bacteria offers interesting alternatives and advantages compared to traditional analytical methods and a variety of bacteriopahge based approaches are described by Schmelcher and Loessner.[20]

With over 5,000 phage morphologically characterized and grouped to a susceptible bacterial host (such as a pathogen), there exists an enormous cache of bacterial-specific sensors that has recently been incorporated into novel bio-recognition assays with heightened sensitivity, specificity, and speed.[21] The capacity of phages to infect and lyse their host cells is utilized in assays which detect the release of cytoplasmic molecules.[20] These assays take many forms, ranging from straightforward visualization of labeled phage as they attach to their specific bacterial hosts, to reporter phage that genetically deposit trackable signals within their bacterial hosts, or the detection of progeny phage or other uniquely identifiable elements released from infected host cells. For example, cell wall recognition, phage adsorption, and injection of phage DNA into the host bacterium have been exploited by detection methods, including capture of cells by immobilized phages and labeling of target organisms by fluorescently tagged phage particles.[20] A comprehensive review of these and other phage-based detection assays, as directed towards the detection and monitoring of bacterial pathogens, is provided by Ripp.[21] The principles and current standing of these approaches along with the application of bacteriophages for detection and control of food borne pathogens have also been summarized by Hagens and Loessner.[22] The advantages of speed, combined with the ability to differentiate between living and dead cells, means that phage-based assays are highly suitable to augment more traditional methods.[22] It is likely that the prevalence and acceptance of phage-based detection methods will increase and that commercial applications will become more widely available and make a valuable contribution to food safety and public health.[22]

3.2 Detection System using Mammalian Cells

A mammalian cell-based biosensor has been reported for application in food defence and food safety.[23] Three prototypes of the biosensor capable of handling different sample types were validated with food and beverages. The biosensor detects pathogenic bacteria interaction with mammalian cells and is able to distinguish pathogenic from non-pathogenic and active from inactive toxins, rendering accurate estimation of the risk associated with these agents.[23] For example, the mammalian B lymphocyte Ped-2E9 cell-line was encapsulated in a collagen matrix and worked as the sensing agent providing positive signal for a broad range of bacterial pathogens such as *Listeria monocytogenes*, enterotoxigenic *Bacillus*, *Vibrio*, *Micrococcus* and *Serratia*, and toxins; alpha-hemolysin from *Staphylococcus aureus*, phospholipase C from *Clostridium perfringens*, cytolysin from sea anemone *Stoichactis helianthus*, listeriolysin O from *L. monocytogenes*, and enterotoxin from *Bacillus*.[23] The detection limit for toxins was found to be 10-40 ng in 2 h, whilst for a model bacterial pathogen, *L. monocytogenes*, detection limits of 10^3-10^4 colony forming units (CFU)/ml were found in 4-6 h, even in the presence of a mixture of higher concentrations of non-pathogenic species of the same genera.[23] The developed prototype biosensors are capable of handling different sample types, such as food and beverages. The data presented by the authors is promising evidence for possible application of this biosensor for rapid detection of multiple pathogens or toxins for food defence and food safety applications.[23] Although the current detection limit is not as low as we expect in drinking water samples, the technique could be tested on concentrated water samples.

3.3 Detection Systems using Surface Modified Biomaterials

Several toxins and pathogens use glycans that cover the surface of mammalian cells to bind and infect the cell. Using a versatile modular synthetic strategy, biotinylated bi- and tetra antennary glyco-conjugates have been developed to capture and detect *Escherichia coli*.[24] In this case, magnetic beads were coated with biotinylated glyco-conjugate or an antibody (for comparison), and these beads were used to capture, isolate, and quantify the recovery of *E. coli.* The authors reported that glycoconjugate-coated magnetic beads out performed antibody-coated magnetic beads in sensitivity and selectivity when compared under identical experimental conditions.[24] Glyco-conjugates could also capture *E. coli* from stagnant water, and the ability of a panel of glyco-conjugates to capture a selection of pathogenic bacteria was also evaluated.[24] The authors reported that the glycoconjugates were found to be very stable, inexpensive and the efficiency of bacterial capture was greater with glycoconjugates as compared to biotiniylated antibodies.[24] These results are expected to lead to an increased interest in developing glyco-conjugate-based high affinity reagents for diagnostics in a variety of different environments. However, this needs further research work to develop glycol-conjugates against each pathogenic strain before adapting this technique to commercial requirements.

4 MOLECULAR MICROBIOLOGICAL BASED DETECTION TECHNIQUES

In water samples, it is difficult precisely to define the size and diversity of the bacterial population. Fortunately, however, novel techniques, based on molecular microbiological methods, are being developed to detect even the scarce presence of pathogens in water directly or indirectly. The range of methods available for molecular detection techniques has also increased in last decade and a significant and up-to-date review of various integrated approaches for bacterial detection has been published.[25 16].

Modified molecular detection techniques generally fall into two main areas that are based on the detection of nucleic acid (DNA / RNA) or specific antigens (immunological based). The nucleic acid based detections lead to the detection of pathogens in water indirectly by reading the genetic alphabets (nucleotide bases). Whereas immunological based detection identifies the signatures in the form of specific antigens coded individually by each organism. Therefore, these two main groups are discussed in 4.1 and 4.2 below.

4.1 Nucleic Acid Based Detection Methods

Techniques that utilize DNA directly allow more powers of discrimination and are increasing in use. Advanced molecular detection methods, based on nucleic acid detection, comprise of some common analytical steps.[27] These include sampling and concentration of the bacteria, extraction of the nucleic acids followed by amplification of the nucleic acids using polymerase chain reaction (PCR), and then detection by various readout technologies for the assessment and characterization of the pathogen. As a consequence, nucleic acid based detection methods are based on discrete samples of the water rather than continuous monitoring. In this section, we have incorporated some molecular techniques that have been used directly or indirectly for the identification of pathogens in environmental samples as we discuss the core basic steps.

4.1.1 Sample Concentration.

To allow fast and efficient concentration and separation of bacterial cells from environmental samples, different procedures have been developed based on several chemical, physical and biological principles. Generally, physical (e.g. Filtration[28] and Ultra centrifugal concentration) or biological-based methods, such as adsorption of bacteria to magnetic beads coated with carboxylated polyvinyl alcohol are adapted for this purpose. Once bacterial cells are concentrated, they may be lysed for their nucleic acid extraction.[29]

4.1.2 Bacterial Nucleic Acid Extraction.
The microbial cellular fraction is separated from the environmental sample as above and then lysed to release the DNA. DNA extraction is carried out by mechanical treatment (e.g. boiling followed by heat cell lysis & bead beating) followed by chemical treatment that includes the use of a surfactant such as sodium dodecyl sulfate buffer, or a surfactant in combination with other chemicals, (e.g. a combination of EDTA, Tris–HCl and Triton) that will dissociate the hydrophobic material of cell membranes.[30]

The challenge of extracting DNA of good quality and quantity from various sources is always present. For example, the critical factors that limit the accuracy and sensitivity of molecular studies for the DNA extracted from the faecal matrix have been reviewed and compared.[30] However, in comparison, DNA extraction from drinking water is likely to be contaminated with less potential inhibitors that may cause problems for the subsequent analysis steps.[30]

4.1.3 Polymerase Chain Reaction (PCR).
Since the concentration of pathogens in water, especially in drinking water, is very low, it is essential to amplify the extracted nucleic acids before its characterization using molecular techniques. Therefore nucleic-acid amplification procedures, by polymerase chain reaction (PCR), remain central to these techniques. PCR remains the gold standard for the identification of many waterborne pathogens and is becoming recognized by official administrations as an acceptable method for the detection of pathogens.[1] It is a sensitive, specific, rapid, and quantitative analytical tool which works by amplifying extracted DNA to allow pathogen detection. To expand the utility of PCR to detect and identify amplified DNA in a multiplex assay, a series of genus- and species-specific primers can be designed and used. It is possible to use multiple primer sets in the same reaction and hence gain additional information, however since such reactions are competitive and affected by multiple factors, the number of primer sets (and hence targets) available is limited.

Many variants of PCR have been developed over the years and reverse transcription polymerase chain reaction (RT-PCR) is a laboratory technique commonly used in molecular biology to generate many copies of a DNA sequence. In RT-PCR, an RNA strand is first reverse transcribed into its DNA complement (complementary DNA, or cDNA) using the enzyme reverse transcriptase and the resulting cDNA is amplified using traditional PCR or real-time PCR (which allows quantification). PCR and RT-PCR have been used for fast (3-5 hour) detection of *salmonellae* and *Vibrio cholerae* in stool, food and environmental water samples.[31 32] Real-time PCR assays have a high sensitivity and allow rapid quantification of pathogens as well as identification e.g. quantification of non-tuberculous mycobacterial DNA in hospital drinking water samples.[33]

A recently developed Recombinase Polymerase Amplification (RPA) technique achieves exponential amplification of DNA without any need for pretreatment of sample.[34] Reactions are sensitive, specific, and rapid and operate at a constant low temperature. The idea of PCR based amplification of DNA by lab-on-chip systems is not new, however, this is the first time a self-sufficient lab-on-a-foil system has been demonstrated for the fully automated analysis of nucleic acids.[35] The system consists of a novel, foil-based centrifugal microfluidic cartridge including pre-stored liquid and dry reagents, and a commercially available centrifugal analyzer for incubation at 37°C and real-time fluorescence detection. The system was characterized with an assay for the detection of the antibiotic resistance gene mecA of *Staphylococcus aureus*. The limit of detection was <10 copies and time-to-result was <20 min. One of the exciting prospects of this RPA based lab-on-a-foil technology is the potential to be used outside a standard molecular microbiological laboratory.

4.1.4 Readout Technologies for Identification.

Diagnosis and pathogen characterization rely increasingly on PCR-based approaches and often PCR is combined with genome sequencing in order to achieve additional resolution. In addition to sequencing, many other feasible and affordable read-out technologies also allow the identification of the presence of a pathogen in a sample, e.g. restriction fragment length polymorphism (RFLP), denaturing gradient gel electrophoresis (DGGE) or temperature gradient gel electrophoresis (TGGE), DNA micro array and spoligotyping. However, robust, reliable and sensitive methods are still required for dissecting pathogenesis, epidemiology, transmission routes and sources of infections.[36]

4.1.4.1 Restriction fragment length polymorphism (RFLP)

RFLP is a PCR based technique of mapping genes with reference to markers for known RFLP loci. In 1993, it was proposed for the designation of transposable elements IS*6110* in *Mycobacterium tuberculosis* for use in strain identification.[37] RFLP typing, with insertion element IS*6110* as a probe, has become the most widely used method for differentiating strains. RFLP analysis was also the basis for early methods of genetic fingerprinting and has proven useful in the identification of samples retrieved from crime scenes, in the determination of paternity, and in the characterization of genetic diversity or breeding patterns in animal populations.[38] RFLP is still a technique used in marker assisted selection. Terminal restriction fragment length polymorphism (TRFLP or sometimes T-RFLP) is a molecular biology technique initially developed for characterizing bacterial communities in mixed-species samples.[38] It works by PCR amplification of DNA using primer pairs that have been labeled with fluorescent tags. The PCR products are then digested using RFLP enzymes and the resulting patterns visualized. The results are analyzed either by simply counting and comparing bands or peaks in the TRFLP profile, or by matching bands from one or more TRFLP runs to a database of known species.[38] The technique is similar in some aspects to DGGE or TGGE. The technique has also been applied to other groups including fungi.[39] The technique itself requires large quantities of DNA and is also relatively labour intensive and, thus, costly.

4.1.4.2 Mass Spectroscopy.

Mass spectrometry (MS) is an analytical technique that is used for determining masses of particles. The MS principle consists of ionizing chemical compounds to generate charged molecules or molecule fragments and measuring their mass-to-charge ratios. A mini review has been published[40] that describe approaches for rapid identification of amoebae including mass spectrometry based detection methods in biological samples. A further valuable use of MS in determining pathogens in healthcare samples is described in section 4.1.4.3.

4.1.4.3. Spoligotyping

Spoligotyping is also a PCR based typing system originally developed for differentiation of strains of *Mycobacterium tuberculosis*. Ishino and colleagues in 1987 found 14 repeats of 29 base pairs (gene sequence) that were interspersed by 32-33 base pair non-repeating spacer sequences adjacent to a gene in *E. coli*. The accumulation of sequenced microbial genomes over the years has allowed genome-wide computational searches and now it has been established that these repeats are found in many bacterial and archaeal genomes.[41 42]

Attempts have, therefore, been made to exploit this information of direct repeat (DR) and spacer sequences for strain identification. Currently these assays are performed by one of two methods.[43] The first, which is the gold standard method, established about 10 years ago, involves the detection of the spacer sequences through amplification of the DR region and subsequent hybridization of the resulting PCR products to nylon membranes with immobilized synthetic probes specific to each spacer. The presence of the spacers is detected by chemilluminescent staining of the nylon membrane.

A second method replaces the time-consuming membrane step, and utilizes a high-throughput, multianalyte flow system (Luminex) that permits the analysis of higher numbers of strains.[44] The synthetic spacer probes are immobilized on microspheres, and detection is achieved via fluorochromes attached to the beads and hybridized PCR product. This method allows the simultaneous analysis of 96 samples, as opposed to 45 samples for the membrane-based method, while producing a digital output that is readily portable.

A new automated primer extension assay has recently been established and successfully validated for the routine typing of *Mycobacterium tuberculosis* complex strains in the research and public health laboratory environments.[45] This multiplexed primer extension-based PCR method adopted the use of mass spectrometry for detection and replaced the hybridization step. This technique has further improved the nylon membrane assay with respect to reproducibility, throughput, process flow, ease of use and data analysis.[45] Hence spoligotyping has many advantages over standard typing procedures such as ease of use use, speed and reliable with very small quantities of DNA (10ng) required and can be easily adapted for waterborne pathogen detection. However, the initial cost to set up a laboratory with the specialist equipment (e.g. mass spectrometers) is high. The level of expertise required would also be very high to successfully analyze samples and interpret the results.

4.1.4.4 DNA Microarray.

The core principle of this technique is hybridization between two complementary nucleic acid sequences. Only labeled target sequences, that bind to a probe sequence, generate a signal. There are a variety of options for making microarrays and obtaining microarray data.[46] Microarray technology has undergone a rapid evolution [47] and an abundance of equipment and reagents have now become available and affordable to a large cross section of the scientific community.

The DNA microarray assay has emerged as a promising alternative for environmental pathogen monitoring. The use of DNA microarrays for comprehensive RNA expression analysis has caused a great deal of interest. Bacterial pathogens have been detected using microarray equipped with probes targeting 16S rRNA sequences.[48 49] Probes have been designed against 38 species, 4 genera, and 1 family of pathogens. The detection sensitivity of the microarray for a waterborne pathogen *Aeromonas hydrophila* was 10^3 cells per sample.[49]

This result demonstrates the effectiveness of the DNA microarray as a semi-quantitative, high throughput pathogen monitoring tool for municipal wastewater.

To meet the demand of rapid and parallel detection and identification of many common pathogenic bacteria, in one experiment a new approach based on gene-chip (microarray) technology has been proposed.[50] Advantages of gene chip technology include miniaturization, high performance, parallelism and automation. One hundred and seventy strains of bacteria in pure culture belonging to 11 genera were discriminated under comparatively similar conditions, and a series of specific hybridization maps corresponding to each kind of bacteria were reported.[50]

Recent studies and advances in microarrays are increasing our capability of detecting hundreds and even thousands of DNA sequences simultaneously and rapidly. With the current progress in micro-fluidics and opto-electronics, the ability to automate a detection/identification system is now being realized. For example, the Lawrence Livermore Microbial Detection Array (LLMDA),[51] is able to detect viruses and bacteria with the use of probes that fit into a checkerboard pattern in the middle of a one-inch wide, three-inch long glass slide. It is also claimed that the current operational version of the LLMDA contains probes that can detect more than 2,000 viruses and about 900 bacteria at the same time.

4.2 Antigen (Immunological) Based Detection Methods

An antigen is a foreign substance that stimulates the production of an antibody when introduced into a living organism.[52] The antibody's capability to distinguish and bind with high affinity to a specific antigen is therefore exploited for qualitative and quantitative measurement of the antigen. The detection signal can be radioactive, colorimetric or fluorescent. The sensitivity and specificity of immunoassays is dependent on the choice of antibodies.

A rapid assay for enterococci was developed using a combined ultrafiltration-biosensor procedure. This study examined the efficacy of a high-volume, hollow fibre ultrafiltration, coupled to biosensor detection for enterococci in surface waters designed to allow same-day public notification of poor water quality.[53] Enterococci were detected when concentrations in the ambient water exceeded the regulatory standard for a single sample ($\geq 10^5$ CFU/100 ml). Rapid detection of enterococci by ultrafiltration, secondary concentration and biosensor analysis demonstrates its potential usefulness for water quality monitoring. The combined procedure required 2.5 h for detection compared with 24 h for EPA method 1600.[53] [54]

An easy-to-use two dot visual filter assay for translation into current water testing in public health laboratories to detect *E. coli* O157:H7 (an enterohemorrhagic bacteria that can cause life-threatening water-borne infections) in water samples has been reported.[55] In a 5 h assay, approximately 1 CFU and approximately 5 CFU of *E. coli* O157:H7 could be detected in 100 ml of water or juice and lake samples respectively. A simple homo-sandwich enrichment strategy could also be used to detect low levels of other water-borne pathogens. In this prototype filter dot assay, anti-*E. coli* O157:H7 monoclonal antibodies were bound to a 0.2 micron antibody coated nitrocellulose filter disk. A 100 ml water sample, spiked with 1-50 CFU of *E. coli* O157:H7, either in the presence or absence of other non-specific bacteria, were filtered for capture on the antibody coated nitrocellulose disk and detection of the pathogen was successfully accomplished.[55]

The costs involved in such molecular detection techniques have decreased over time. More recent improvements in detection techniques have also allowed the simultaneous detection of multiple targets in a single assay. However, it is still believed that sample processing and molecular detection techniques need to be further advanced to allow rapid and specific identification of microbes of public health concern from complex environmental samples.[26] Consideration of the more recent techniques however, suggests that it may be time to reassess these alternatives molecular techniques that show promise for the water industry.

5 CONCLUSIONS

Within research laboratories worldwide there are multiple methods being combined and developed to improve the sensitivity and reliability of techniques for pathogen detection. Many viable laboratory methods for pathogen detection exist now and more are in development.[16] This review provides an overview of some potential culture independent methods that may be appropriate for use by the water industry for pathogen detection. In some cases, examples are also given of techniques that have been used for pathogen detection in healthcare applications as this provides an added source of potential techniques for drinking water pathogen detection.

Technologies are developing fast and numerous techniques have the potential for detecting microorganisms in samples in a shorter time and with a greater precision than culture based techniques. However, at present the culture independent methods are predominately laboratory based, require specialist equipment and knowledge, remain reliant on discrete sampling, and need a high degree of sample pre- and post-processing.

Technological challenges remain that limit the application of these new technologies, currently prohibiting transfer from the research lab into the field. It is possible to detect pathogens using culture independent techniques already. However, as the technology matures, the application of new techniques will also require key decisions to be made by the water industry and regulators in terms of deciding the balance between level and type of information required, time and complexity of analysis and costs.

Acknowledgements

This authors wish to thank the UK Engineering and Physical Sciences Council (EPSRC) for the provision of funding (EP/H023488/1, EP/E053556/01 and GR/S84347/01)

References

1. M. Bouzid, D. Steverding and K. M. Tyler, *Curr Opin Biotechnol*, 2008, **19**, 302.
2. E. C. Environment. *The Drinking Water Directive (DWD), Council Directive 98/83/EC,* 1998; Available from: http://ec.europa.eu/environment/water/water-drink/index_en.html (accessed 28th July 2011).

3. E. C. Environment. *Revision of Directive 98/83/EC (Revised Draft of Drinking Water Directive (DWD))*, 2008; Available from: http://circa.europa.eu/Public/irc/env/drinking_water_rev/library?l=/working_revision/distributionpdf/_EN_1.0_&a=d (accessed 28th July 2011).
4. WHO. *Guidelines for Drinking Water Quality. Vol 1 3rd Ed.* 2004; Available from: http://www.who.int/water_sanitation_health/dwq/GDWQ2004web.pdf (accessed 28th July 2011).
5. E. C. Environment. *Revision of the Drinking Water Directive of the draft minutes of the meetings of 2nd and 3rd Expert Group on Microbiology held on 17/18 October 2007 and on 27/28 February 2008 at the JRC Ispra, Italy.* 2008 15/04/2008; Available from: http://circa.europa.eu/Public/irc/env/drinking_water_rev/library?l=/microbiological/17102007_28022008/_EN_1.0_&a=d (accessed 28th July 2011).
6. Y. Qin, H. Wu, X. Xiao, Y. Yu, D. Liu, X. Li and Y. Tang, *Sheng Wu Gong Cheng Xue Bao*, 2009, **25**, 1497.
7. USEPA. *Literature review of molecular methods for simultaneous detection of pathogens in water.* 2008; Available from: http://www.epa.gov/nrmrl/pubs/600r07128/600r07128.pdf (accessed 28th July 2011).
8. S. Tajbakhsh, S. Gharibi, K. Zandi, R. Yaghobi and G. Asayesh, *Eur Rev Med Pharmacol Sci*, 2011, **15**, 313.
9. G. Vesey, N. Ashbolt, E. J. Fricker, D. Deere, K. L. Williams, D. A. Veal and M. Dorsch, *J Appl Microbiol*, 1998, **85**, 429.
10. B. Nehme, M. Henry and D. Mouginot, *J Neurosci Methods*, 2011, **196**, 281.
11. D. N. King, K. P. Brenner and M. R. Rodgers, *J Water Health*, 2007, **5**, 295.
12. A. Pianetti, A. Manti, P. Boi, B. Citterio, L. Sabatini, S. Papa, M. B. Rocchi and F. Bruscolini, *Int J Food Microbiol*, 2008, **127**, 252.
13. H. A. Keserue, H. P. Fuchslin and T. Egli, *Appl Environ Microbiol*, 2011.
14. J. Barbosa, C. Bragada, S. Costa-de-Oliveira, E. Ricardo, A. G. Rodrigues and C. Pina-Vaz, *Eur J Clin Microbiol Infect Dis*, 2010, **29**, 1147.
15. S. C. Bendall, E. F. Simonds, P. Qiu, A. D. Amir el, P. O. Krutzik, R. Finck, R. V. Bruggner, R. Melamed, A. Trejo, O. I. Ornatsky, R. S. Balderas, S. K. Plevritis, K. Sachs, D. Pe'er, S. D. Tanner and G. P. Nolan, *Science*, 2011, **332**, 687.
16. M. Zourob, S. Elwary and A. P. F. Turner, Springer, New York ; London, 2008, pp. xxxii, 970 p.
17. K. F. Reardon, Z. Zhong and K. L. Lear, *Adv Biochem Eng Biotechnol*, 2010, **116**, 99.
18. E. Alocilja and Z. Muhammad-Tahir, in *Principles of bacterial detection biosensors, recognition receptors and microsystems*, eds. M. Zourob, S. Elwary and A. P. F. Turner,
19. A. K. Singh, D. Senapati, S. Wang, J. Griffin, A. Neely, P. Candice, K. M. Naylor, B. Varisli, J. R. Kalluri and P. C. Ray, *ACS Nano*, 2009, **3**, 1906.

Springer, New York ; London, 2008, pp. 377.

20. M. Schmelcher and M. J. Loessner, in *Principles of bacterial detection biosensors, recognition receptors and microsystems*, eds. M. Zourob, S. Elwary and A. P. F. Turner, Springer, New York ; London, 2008, pp. 731.
21. S. Ripp, *Adv Biochem Eng Biotechnol*, 2010, **118**, 65.
22. S. Hagens and M. J. Loessner, *Appl Microbiol Biotechnol*, 2007, **76**, 513.
23. P. Banerjee and A. K. Bhunia, *Biosens Bioelectron*, 2010, **26**, 99.
24. D. M. Hatch, A. A. Weiss, R. R. Kale and S. S. Iyer, *Chembiochem*, 2008, **9**, 2433.

25. K. A. Gilbride, D. Y. Lee and L. A. Beaudette, *J Microbiol Methods*, 2006, **66**, 1.
26. J. R. Stewart, R. J. Gast, R. S. Fujioka, H. M. Solo-Gabriele, J. S. Meschke, L. A. Amaral-Zettler, E. Del Castillo, M. F. Polz, T. K. Collier, M. S. Strom, C. D. Sinigalliano, P. D. Moeller and A. F. Holland, *Environ Health*, 2008, **7 Suppl 2**, S3.
27. I. Brettar and M. G. Hofle, *Curr Opin Biotechnol*, 2008, **19**, 274.
28. Y. Wang, F. Hammes, N. Boon and T. Egli, *Environ Sci Technol*, 2007, **41**, 7080.
29. M. G. Weinbauer, I. Fritz, D. F. Wenderoth and M. G. Hofle, *Appl Environ Microbiol*, 2002, **68**, 1082.
30. D. Rapp, *J Appl Microbiol*, 2010, **108**, 1485.
31. A. M. Silva, H. Vieira, N. Martins, A. T. Granja, M. J. Vale and F. F. Vale, *J Appl Microbiol*, 2010, **108**, 1023.
32. J. M. Ling, *Hong Kong Med J*, 2009, **15 Suppl 2**, 26.
33. Z. Hussein, O. Landt, B. Wirths and N. Wellinghausen, *Int J Med Microbiol*, 2009, **299**, 281.
34. O. Piepenburg, C. H. Williams, D. L. Stemple and N. A. Armes, *PLoS Biol*, 2006, **4**, e204.
35. S. Lutz, P. Weber, M. Focke, B. Faltin, J. Hoffmann, C. Muller, D. Mark, G. Roth, P. Munday, N. Armes, O. Piepenburg, R. Zengerle and F. von Stetten, *Lab Chip*, 2010, **10**, 887.
36. A. Curry and H. V. Smith, *Parasitology*, 1998, **117 Suppl**, S143.
37. J. D. van Embden, M. D. Cave, J. T. Crawford, J. W. Dale, K. D. Eisenach, B. Gicquel, P. Hermans, C. Martin, R. McAdam, T. M. Shinnick and et al., *J Clin Microbiol*, 1993, **31**, 406.
38. Wikipedia. *Restriction fragment length polymorphism*. 2010; Available from: http://en.wikipedia.org/wiki/Restriction_fragment_length_polymorphism (accessed 28th July 2011).
39. A. Szekely, R. Sipos, B. Berta, B. Vajna, C. Hajdu and K. Marialigeti, *Microbial Ecol*, 2009, **57**, 522.
40. G. S. Visvesvara, H. Moura and F. L. Schuster, *FEMS Immunol Med Microbiol*, 2007, **50**, 1.
41. V. Kunin, R. Sorek and P. Hugenholtz, *Genome Biol*, 2007, **8**, R61.
42. I. Grissa, G. Vergnaud and C. Pourcel, *BMC Bioinformatics*, 2007, **8**, 172.
43. J. R. Driscoll, *Methods Mol Biol*, 2009, **551**, 117.
44. L. S. Cowan, L. Diem, M. C. Brake and J. T. Crawford, *J Clin Microbiol*, 2004, **42**, 474.
45. C. Honisch, M. Mosko, C. Arnold, S. E. Gharbia, R. Diel and S. Niemann, *Journal of Clinical Microbiology*, 2010, **48**, 1520.
46. V. G. Cheung, M. Morley, F. Aguilar, A. Massimi, R. Kucherlapati and G. Childs, *Nat Genet*, 1999, **21**, 15.
47. A. J. Holloway, R. K. van Laar, R. W. Tothill and D. D. Bowtell, *Nat Genet*, 2002, **32** Suppl, 481.
48. D. D. Bowtell, *Nat Genet*, 1999, **21**, 25.
49. D. Y. Lee, H. Lauder, H. Cruwys, P. Falletta and L. A. Beaudette, *Sci Total Environ*, 2008, **398**, 203.
50. L. Q. Jin, J. W. Li, S. Q. Wang, F. H. Chao, X. W. Wang and Z. Q. Yuan, *World J Gastroenterol*, 2005, **11**, 7615.

51. LLNL. *New LLNL detection technology identifies bacteria, viruses, other organisms within 24 hours (News release).* May 2010; Available from: https://publicaffairs.llnl.gov/news/news_releases/2010/NR-10-05-02.html (accessed 28th July 2011)
52 Oxford English Dictionary http://www.oed.com/ (accessed 28th July 2011)
53. S. D. Leskinen and D. V. Lim, *Appl Environ Microbiol*, 2008, **74**, 4792.
54. S. D. Leskinen, V. J. Harwood and D. V. Lim, *J Water Health*, 2009, **7**, 674.
55. S. Kamma, L. Tang, K. Leung, E. Ashton, N. Newman and M. R. Suresh, *J Immunol Methods*, 2008, **336**, 159.

DETECTION OF FAECAL CONTAMINATION IN THE DRINKING WATER OF SMALL COMMUNITY WATER SUPPLY PLANTS IN FINLAND

Tarja Pitkänen[1], Helvi Heinonen-Tanski[2], Marja-Liisa Hänninen[3] and Ilkka T. Miettinen[1]

[1]National Institute for Health and Welfare, P.O. Box 95, FI-70701 Kuopio, Finland
[2]University of Eastern Finland, P.O. Box 1627, FI-70211 Kuopio, Finland
[3]University of Helsinki, P. O. Box 57, FI-00014 Helsinki, Finland

1 INTRODUCTION

Safe drinking water is essential for human health and contaminated drinking water has the potential to produce serious health concerns[1]. The contamination may take place at the water source (reservoir), at the water treatment plant or within the distribution system. Human or animal faeces may contain pathogenic microbes (bacteria, viruses, protozoans, helminths). Thus, waterborne infections are highly probable in cases when there are failures to prevent faecal contamination of drinking water. Since the drinking water distribution networks usually serve a large number of individuals, the faecal contamination of water may lead to a wide-spread outbreaks of disease.

Microbial contamination of water can be traced primarily to the presence of animal or human faeces that may originate from untreated sewage entering the distribution system, from animal waste being carried by rain runoff or by melting snow, or from failure or breakdown in the water treatment process[2, 3]. Microbiological drinking water safety is based on the prevention of access of faecal material to water, and purification and disinfection treatments at the waterworks.

In spite of these measures, waterborne outbreaks occur even in industrialized countries with centralized water supplies and sanitation. The majority of such outbreaks take place at relatively small groundwater abstraction plants utilizing minimal water treatment, and the source of contamination often remains unclear. There is an evident need for more specific knowledge of the environmental hazards that endanger water safety and cause contaminations. The accuracy and speed of detection and source tracking of the microbial hazards is often compromised by inadequate sampling and unsatisfactory methods employed[4]. Especially at small water supply plants, the sampling and analysis is performed only infrequently. The microbiological methods currently applied in water quality monitoring are often laborious and time-consuming. Regulatory water quality requirements, in the future, should promote microbial hazard identification, determination of the acceptable level of risk and adoption of risk management practices[1].

2 DETECTION METHODS OF FAECAL INDICATOR BACTERIA

In the prevention of waterborne human diseases, it is important to have accurate and reliable methods for determining the microbial water quality. Monitoring of the hygienic quality of drinking water is based on the indicator organism approach, since many of the known waterborne pathogens are difficult to detect and it is not feasible to issue regulations requiring the authorities to monitor the complete spectrum of microorganisms[5]. Bacterial indicator organisms are required for drinking water monitoring purposes and the detection methods for these indicators are standardized[4]. Coliform bacteria, *Escherichia coli*, intestinal enterococci, *Clostridium perfringens* and heterotrophic plate counts (HPC) are utilized around the world as indicator organisms.

The fast and reliable detection and confirmation of microbial drinking water safety is essential for the protection of public health. Currently, the detection of *Escherichia coli* is most commonly utilized as an indicator of water safety since *E. coli* is considered to be the best known and the most practicable indicator of fresh faecal contamination[6, 7]. The detection of *E. coli* is performed at laboratories employing cultivation methods that conventionally take 2-3 working days before a confirmed *E. coli* result is available[8]. There are faster *E. coli* methods based on chromogenic/fluoregenic detection that may give the confirmed result on the following day after the start of incubation.

Several methods for the enumeration of coliform bacteria and *Escherichia coli* were compared in a study with non-disinfected waters[9]. Five alternative cultivation methods for detection and enumeration of *Escherichia coli* and coliform bacteria, indicators of microbial water quality, from non-disinfected waters were compared with the reference method ISO 9308-1 (LTTC). Three of them: Colilert®-18, Chromocult® coliform agar and chromogenic E.coli/coliform medium achieved equal or higher coliform bacteria counts than LTTC, but the alternative media seemed to produce lower *E. coli* counts than LTTC. It was concluded that Colilert®-18, Chromocult® coliform agar and chromogenic *E. coli*/coliform medium can be used as potential alternative media once the specific problems related to false positive and false negative results can be eliminated.

Since several methods are often available for the same purpose, the selection of methods varies in different countries and even in different laboratories within a single country. When the large databases of water quality results are being evaluated, a comparison of the results obtained using different media in different laboratories may be needed. The international standard method ISO 17994 lists the criteria for establishing equivalence between two quantitative microbiological cultivation media[10] and, in recent year,s this standard has streamlined the various procedures used in comparison trials[11].

In Europe, the European Drinking Water Directive lists reference methods that must be used in microbial water quality monitoring for regulatory purposes[12]. However, the directive states that alternative methods may be used if a Member State is able to demonstrate that the results are at least as reliable as those produced by the reference method. This statement has led to numerous comparison trials especially for the detection methods of coliform bacteria and *E. coli* in Europe since the selectivity of the stated reference method[13] is claimed to be inadequate. In Finland, conventional mEndo Agar LES (Les Endo) medium is currently the most commonly used method for *E. coli* and coliform detection from drinking water. However, the Colilert® Quantitray is increasingly being employed by the water industry and in Finnish regulatory monitoring after its acceptance for that purpose; initially for bathing water monitoring but later also for drinking water monitoring[14].

The survival, transport, removal and inactivation of waterborne pathogens, especially certain viruses and protozoa, differ substantially from the survival of conventional indicator bacteria in the aqueous environment[4]. This decreases the predictive value of indicator bacteria for guaranteeing the microbiological safety of water. The presence of faecal indicator bacteria is a sign of faecal contamination, but the absence of the cultivable faecal indicators can not be taken as an assurance of non-contamination. It has been recommended that microbial indicators, more resistant than *E. coli* in environmental conditions, or physico-chemical parameters may be helpful when used in conjunction with *E. coli* to supplement water quality assessment[7]. Intestinal enterococci and *Clostridium perfringens* are more resistant bacterial indicators for faecal pollution[6], especially the spores of *C. perfringens* are extremely persistent in the environment[15] and are also resistant to water disinfection processes[16].

When drinking water samples are tested, even a very low concentration of faecal bacteria needs to be detected. This requires the availability of efficient water concentration techniques[17, 18] even though it has been stated that in compliance monitoring, the 100 ml volumes may be sufficient and a large number of small-volume samples is preferable to a smaller number of large-volume samples[19]. For hazard identification purposes and in cases where there is a suspicion of waterborne illness transmission, however, larger volumes may be essential[20, 21].

3 DRINKING WATER QUALITY IN FINLAND

Surface waters and groundwater serve as sources of potable water for communities. In Finland, 37 % of the drinking water supplied by waterworks originates from surface water, 52 % is groundwater and 11 % is artificially recharged groundwater[22]. The selection of potable water source depends on the local circumstances, local source water quality, and the availability of water from different sources.

There are about 170 large water supplies producing drinking water more than 1 000 m^3/day in Finland serving about 77 % of the population (i.e. 4.1 million consumers). There is no exact number available on the current number of water supplies smaller this value. Based on voluntarily reported water quality data in 2008[23], there were at least 111 medium-size supplies (the volume of water produced 400 $\leq$ 1000 m^3/day), at least 188 small supplies (100 $\leq$ 400 m^3/day) and at least 440 very small supplies (10 $\leq$ 100 m^3/day). The majority of these smaller supplies (93 %) are ground water intake plants often managed by local water cooperatives, financed by charging very operation and maintenance fees, and are operated by people working on the voluntary basis. The groundwater suppliers in Finland often pump the water into their distribution system without treatment or disinfection.

Based on voluntarily reported water quality monitoring results from 2008, the microbiological water quality on large supplies was better than the quality of water produced by smaller water supplies[23]. For example, more than 99.4 %, 98.2 %, 97.0 % and 94.4 % of the very small supplies (producing 10 – 100 m^3/day) did not exceed the quality criterion for *E. coli*, intestinal enterococci, coliform bacteria and heterotrophic plate count, respectively, compared with 99.9 %, 99.7 %, 99.4 % and 99.2 % compliance from the results originating from large drinking water supplies producing water more than 1 000 m^3/day.

4 A SURVEY AT SMALL COMMUNITY WATER SUPPLIES IN FINLAND

In a recent study, the hazards to the microbial safety of drinking water were estimated at small community water supplies in central Finland[24]. A survey was carried out where microbiological and chemical hazards compromising water quality at twenty community water supplies were identified (Table 1). The results showed that faecal indicator bacteria, *Escherichia coli*, intestinal enterococci or *Clostridium perfringens*, could be detected in 10 % of the groundwater samples originating from five water supply plants, all of them serving less than 250 consumers. With respect to the other water quality hazards, coliform bacteria were detected in 40% of the samples using enhanced methods, 54% of the samples were coliphage positive and 16% of the samples contained more than 0.10 mg/l nitrite.

Table 1 *Size, management type and utilization of the disinfection at the studied 20 ground water supply plants in central Finland.*

Water supply size (m^3/d)	Number of supplies	Management (Cooperative/Municipal)	Disinfection (Yes/No)
Large > 1 000	8	1 / 7	6 / 2
Medium-size 400 ≤ 1 000	5	1 / 4	1 / 4
Small 100 ≤ 400	5	5 / 0	1 / 4
Very small 10 ≤ 100	2	2 / 0	0 / 2

5 ON-SITE HAZARDS ASSOCIATING WITH FAECAL CONTAMINATION

In conjunction with the sampling campaigns arranged during the water quality survey[24], on-site hazard identification was conducted at each water supply and a scored multiple-choice questionnaire was completed. On-site hazards to water quality were identified to be present at the all studied 20 community water supplies. The main hazard identified to reduce water safety was insufficient protection against the influence of surface water. The identified characteristics enabling the impact of surface water on groundwater quality were:

i. a poor well construction and maintenance enabling surface water runoffs directly into raw water wells;
ii. an insufficient depth of the protective soil layer above the groundwater table; and
iii. the possibility of uncontrolled river or lake bank infiltration.

The presence of on-site technical hazards to water safety, showed the vulnerability of groundwater used for drinking purposes. Four supplies, out of total five with the observed presence of faecal indicator bacteria, did not utilize disinfection as a preventive measure to avoid waterborne illnesses and, thus, pumped the contaminated water directly into their distribution network.

Later on, the studies on Finnish small community water supplies continued and then, in addition to the findings of faecal indicator bacteria from some samples, *Giardia intestinalis* was detected for the first time from groundwater supplies in Finland[25]. In that study, the main contaminating hazard identified was sewage treatment activities (potentially leaking household septic tanks) located in close proximity to the wells.

6 DISCUSSION

Improved site-specific identification of hazards causing potential health risks is needed, but a low risk estimate does not completely guarantee water safety. In the future, multiple parameter microbial hazard identification, with a wide set of indicators such as *E. coli*, enterococci, coliform bacteria, *C. perfringens* and coliphages analysed from sample volumes of 1000 ml or larger volumes during an intensive sampling period, might raise awareness of the possible health risks at small water supply plants. As a preventive measure, the upgrading of water treatment processes, utilization of disinfection and systematic risk management at small groundwater supplies are recommended.

References

1 WHO. 2011. Guidelines for drinking-water quality. http://whqlibdoc.who.int/publications/2011/9789241548151_eng.pdf: World Health Organization
2 Geldreich EE. 1990. In *Drinking water microbiology: progress and recent developments*, ed. GA McFeters. New York: Springer-Verlag
3 Percival SL, Walker, J.T., Hunter P.R. 2000. *Microbiological Aspects of Biofilms and Drinking Water*. Boca Raton, Florida: CRC Press. 229 pp.
4 Pitkänen T. 2010. *Studies on the detection methods of Campylobacter and faecal indicator bacteria in drinking water*: University of Eastern Finland and National Institute for Health and Welfare. 183 pp.
5 NRC. 2004. *Indicators for Waterborne Pathogens*. Washington, D.C.: The National Academic Press. 315 pp.
6 Edberg SC, Rice EW, Karlin RJ, Allen MJ. 2000. *J Appl Microbiol* **88**: 106S-16S
7 Tallon P, Magajna B, Lofranco C, Leung KT. 2005. *Water Air Soil Poll* **166**: 139-66
8 Rompre A, Servais P, Baudart J, de Roubin MR, Laurent P. 2002. *J Microbiol Meth* **49**: 31-54

9 Pitkänen T, Paakkari P, Miettinen IT, Heinonen-Tanski H, Paulin L, Hänninen ML. 2007. *J Microbiol Meth* **68**: 522-9
10 ISO 17994. 2004. *International Organization for Standardization*: 1-14
11 Sartory DP. 2005. *Water Sa* **31**: 393-6
12 European Union. 1998. *Official J. Eur. Communities* **L330**: 32-54
13 ISO 9308-1. 2000. *International Organization for Standardization*: 1-10
14 Valvira. 2009. Talousveden laatuvaatimuksista ja valvontatutkimuksista annetun sosiaali- ja terveysministeriön asetuksen (461/2000) mukaisessa talousveden viranomaisvalvonnassa sovellettavat koliformisten bakteerien ja *Escherichia coli*-bakteerin määritysmenetelmät [In Finnish]. http://www.valvira.fi/files/Colilert-p%C3%A4%C3%A4t%C3%B6s_080509.pdf: National Supervisory Authority for Welfare and Health
15 Schijven JF, de Bruin HAM, Hassanizadeh SM, Husman AMD. 2003. *Water Res* **37**: 2186-94
16 Hijnen WAM, Beerendonk EF, Medema GJ. 2006. *Water Res* **40**: 3-22
17 Hijnen WAM, Van Veenendaal DA, Van der Speld WMH, Visser A, Hoogenboezem W, Van der Kooij D. 2000. *Water Res* **34**: 1659-65
18 Hill VR, Polaczyk AL, Hahn D, Narayanan J, Cromeans TL, et al. 2005. *Appl Environ Microbiol* **71**: 6878-84
19 Locas A, Barthe C, Margolin AB, Payment P. 2008. *Can J Microbiol* **54**: 472-8
20 Hänninen ML, Haajanen H, Pummi T, Wermundsen K, Katila ML, et al. 2003. *Appl Environ Microbiol* **69**: 1391-6
21 Hargy TM, Rosen J, LeChevallier M, Friedman M, Clancy JL. 2010. *J AWWA* **102**: 79
22 Isomäki E, Valve, M., Kivimäki, A.-L., Lahti, K. 2008. *Operation and maintenance of small waterworks*. Helsinki: Finnish Environment Institute
23 Zacheus O. 2010. *Talousveden valvonta ja laatu vuonna 2008. Yhteenveto viranomaisvalvonnan tuloksista. [In Finnish]*, Helsinki
24 Pitkänen T, Karinen P, Miettinen IT, Lettojarvi H, Heikkila A, et al. 2011. *Ambio* **40**: 377-90
25 Juselius T, Pitkänen T, Isomäki E, Miettinen IT, Valve M, et al. 2011. *In preparation*

MONITORING AND ASSESSMENT IN A WATER TREATMENT PLANT USING BANKFILTRATED RAW WATER IN DUESSELDORF, GERMANY

Vera Schumacher*, Timo Binder, Hans-Peter Rohns, Christoph Wagner

Stadtwerke Duesseldorf AG, Wiedfield 50, 40589 Duesseldorf, GERMANY
*Email: vschumacher@swd-ag.de

1 INTRODUCTION

The Stadtwerke Duesseldorf Company is an energy provider and water supplier. Three waterworks produce drinking water for the cities of Duesseldorf, Mettmann and Erkrath in Germany using riverbank filtered raw water. In order to provide safe drinking water to the consumer, the water passes through a number of treatment steps according to the multiple barrier system. After riverbank filtration the water is ozonated, decarbonated and finally filtered by a two tiered activated carbon filter (Figure 1).

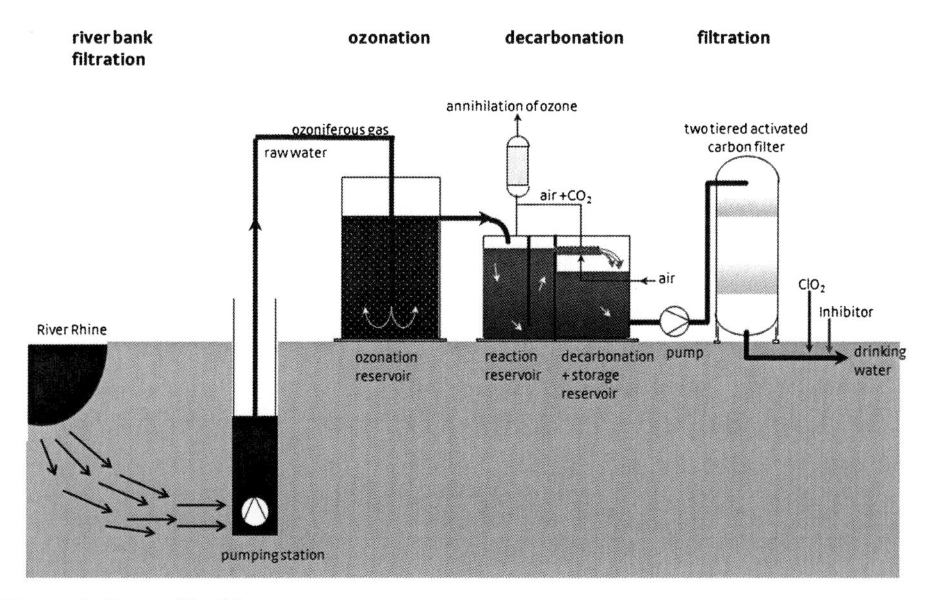

Figure 1: *Duesseldorf Process.*

The whole supply chain (catchment, treatment and distribution) is monitored continuously to reveal possible leakages in the multiple barrier system and to ensure acceptable distributing drinking water quality.
Despite this operating expense coliforms were regularly detected after riverbank filtration in one of the three waterworks of the Stadtwerke Duesseldorf. Coliforms occur naturally in the intestine of humans and animals and therefore are commonly measured as index parameters indicating faecal contamination. In addition, they may act as pathogens and cause infections. Their occurrence in the River Rhine was expected because of its usage as discharge system. But the appearance of coliforms in the river bank filtrate was unusual and led to a catchment survey and a risk assessment according to the Water Safety Plan recommendations of the WHO.

2 METHOD AND RESULTS

2.1 Data research

First of all, a data search was performed. All coliform findings in raw water samples between 1990 and 2005 were analysed and the part of exceeding a special limit was evaluated. In Germany, there is no limit value for coliforms in raw water. Therefore, the threshold for coliforms in drinking water (1 cfu/100 ml) was used for evaluation. Before 2000 over 50 % of the samples exceeded the limit value (except in 1991) (Figure 2).

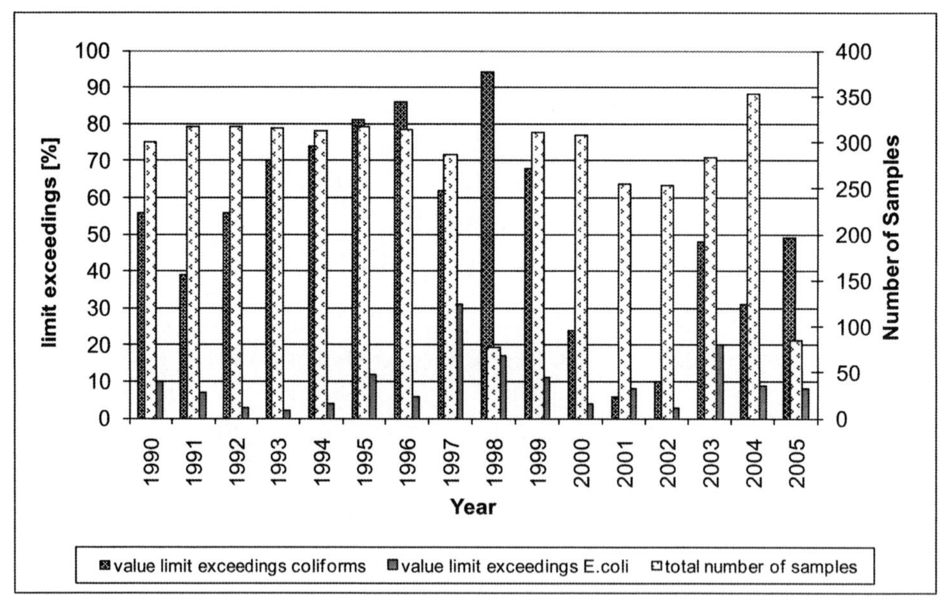

Figure 2: *Limit exceeding compared to the total examined samples in the observation period from 1990 to 2005.*

Except in 1997, when in 30 % of the samples *E.coli* was identified, less than 20 % of the probes contained *E.coli*.
During the observation period the methods used for coliforms and *E.coli* enumeration changed (Table 1).

Table 1: *Comparison of used methods for identifying coliforms and its results.*

Method	Practice Period	Number of samples	Limit Exceeding [%]	
			Coliforms	*E.coli*
membrane filtration on Endo-Agar (according to German drinking water ordinance)	01.01.1990 – 31.08.1999	2921	69	10
membrane filtration on MacConkey-Agar (according to German drinking water ordinance)	01.09.1999 – 31.12.2002	920	15	38
membrane filtration on Lactose-TTC-Agar (according to German drinking water ordinance)	01.01.2003 – 01.08.2005	146	40	38
Colilert®	01.01.2003 – 01.08.2005	508	70	7,5

Depending on the method used, the number of limit exceeding results varies [1,2]. This is one reason for the changing coliform level detected in raw water. And in some cases, the method may be the only reason for the evidence of coliforms.

2.2 Influence of Water level of River Rhine on the occurrence of Coliforms in Raw Water

The water source itself, the River Rhine, could influence the appearance of coliforms in the raw water of the treatment plant. The data search showed large differences in magnitude and incidence of coliform contamination of the raw water depending on the method of analysis (Figures 3 to 6). But even in years with the same method of analysis, there was no direct correlation between water level, or the chang in water level, of the River Rhine and the coliform contamination in raw water (Figures 3 and 4).

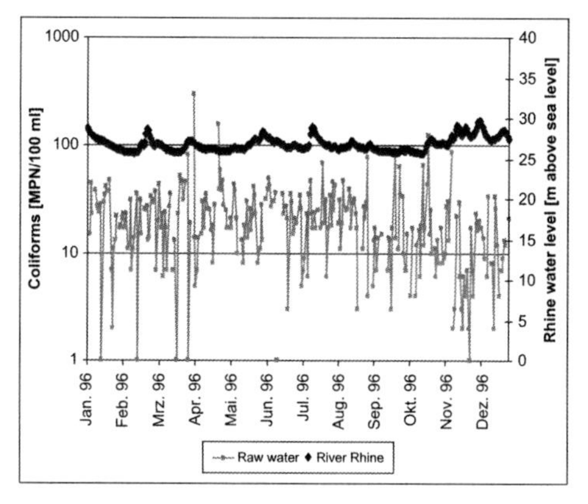

Figure 3: *Coliform contamination of raw water and water level of River Rhine in 1996.*

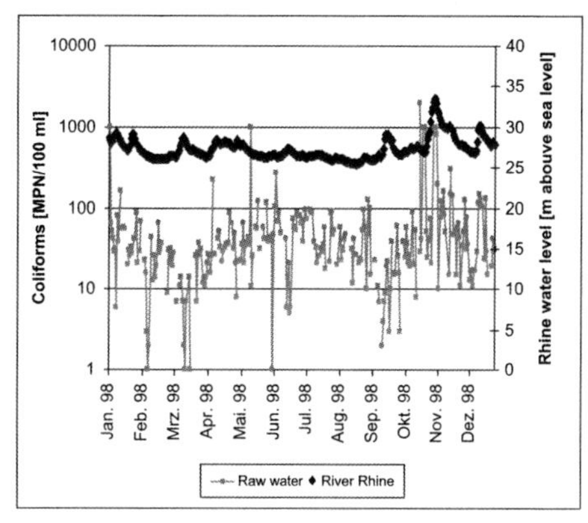

Figure 4: *Coliform contamination of raw water and water level of River Rhine in 1998.*

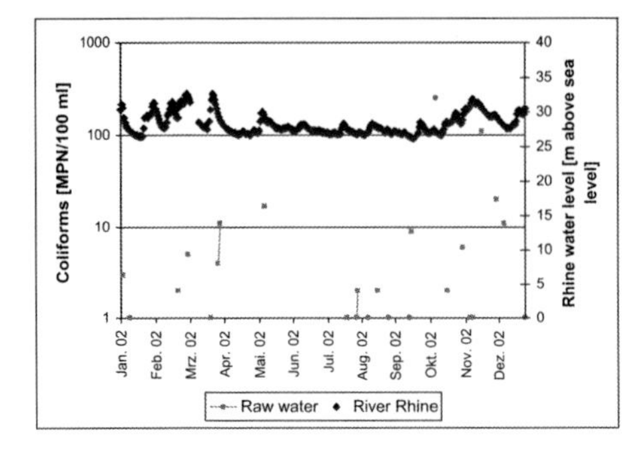

Figure 5: *Coliform contamination of raw water and water level of River Rhine in 2002.*

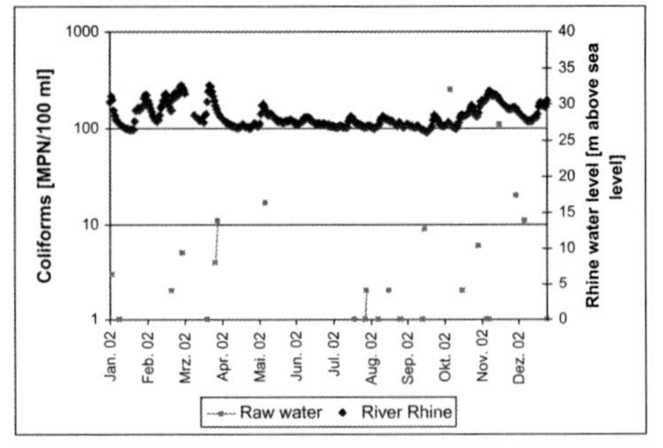

Figure 6: *Coliform contamination of raw water and water level of River Rhine in 2004.*

Further research regarding this topic was accomplished. Therefore the coliform contamination and the contamination with *E.coli* of the River Rhine were compared with the water level (Figure 7). An obvious correlation was detected most of the time during the observation period.

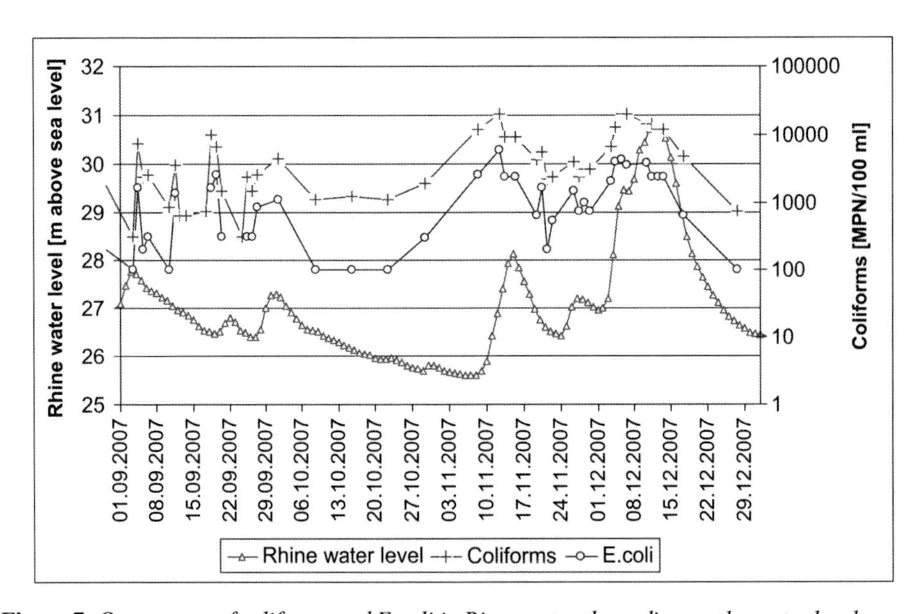

Figure 7: *Occurrence of coliforms and E.coli in River water depending on the water level.*

Before every increase of the water level the contamination - both of coliforms and *E.coli* – also increased. A precise time-relationship cannot be determined because the samples for coliform analysis were only taken from Monday to Thursday each week.
The river bank filtration reduced the number of both - coliforms and *E.coli* - by more than two $\log_{10}$ orders. (Figures 8 and 9).

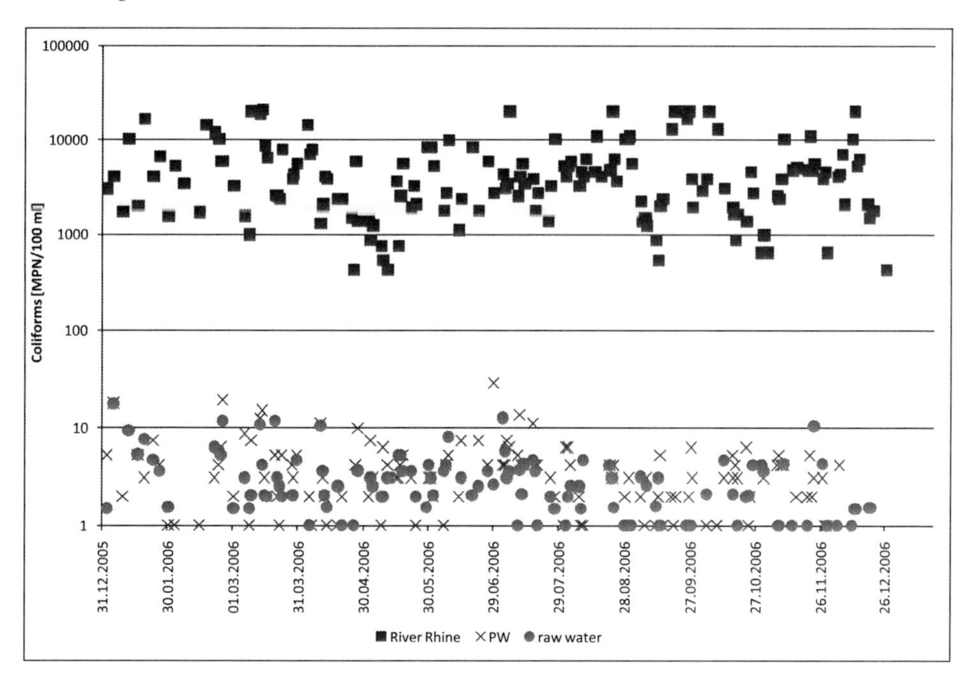

Figure 8: *Coliform contamination of the River Rhine, the raw water of both pumping stations (PW) and water treatment plant (raw water).*

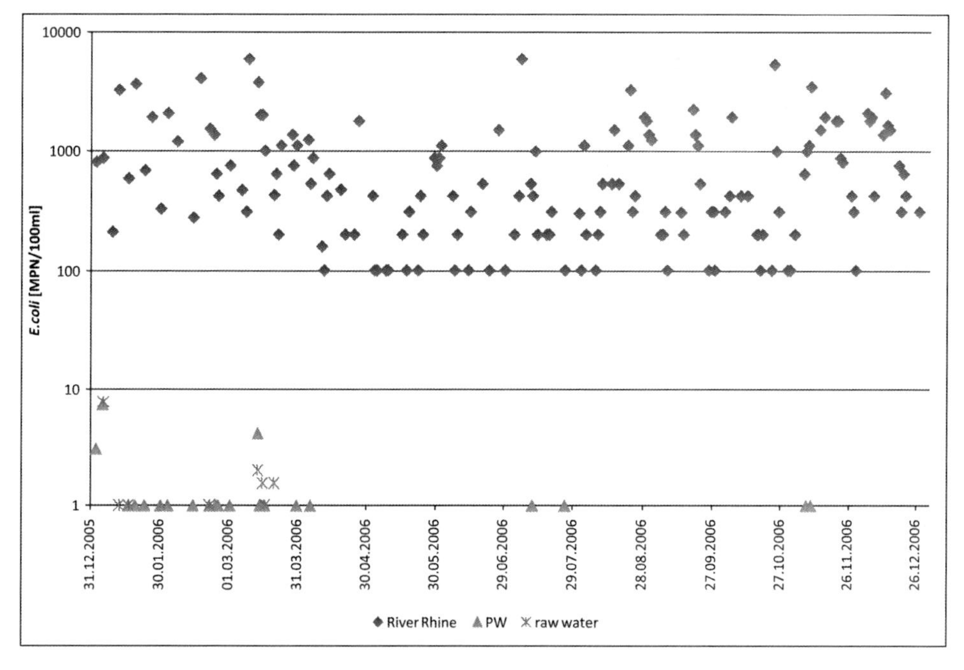

Figure 9: *Contamination of the River Rhine and the raw water of both pumping stations (PW) and water treatment plant (raw water) with E.coli.*

An influence of the coliform contamination of the River Rhine on raw water quality was observed under the conditions outlined below.

2.3 Pumping station

To identify the source of the coliforms the raw water delivered to the water treatment plant was investigated. In addition, the raw water of the three corresponding pumping stations (Figure 10) was examined (Colilert®, 4 days a week).

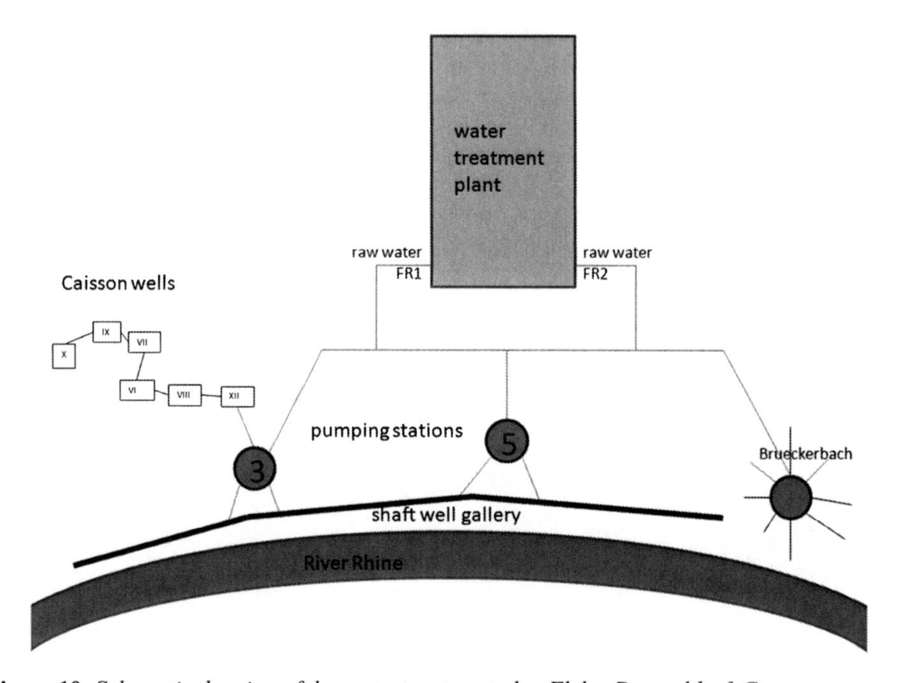

Figure 10: *Schematic drawing of the water treatment plan Flehe, Duesseldorf, Germany.*

Coliforms were detected in two of the pumping stations. But coliforms were only detected in the raw water of pumping station 5, when pumping station 3 was disabled (Figure 11). During this observation, no coliforms were detected in the pumping station Brueckerbach. These results led to the conclusion that the origin of the coliforms was located in the catchment area of pumping station 3 (Figure 10).

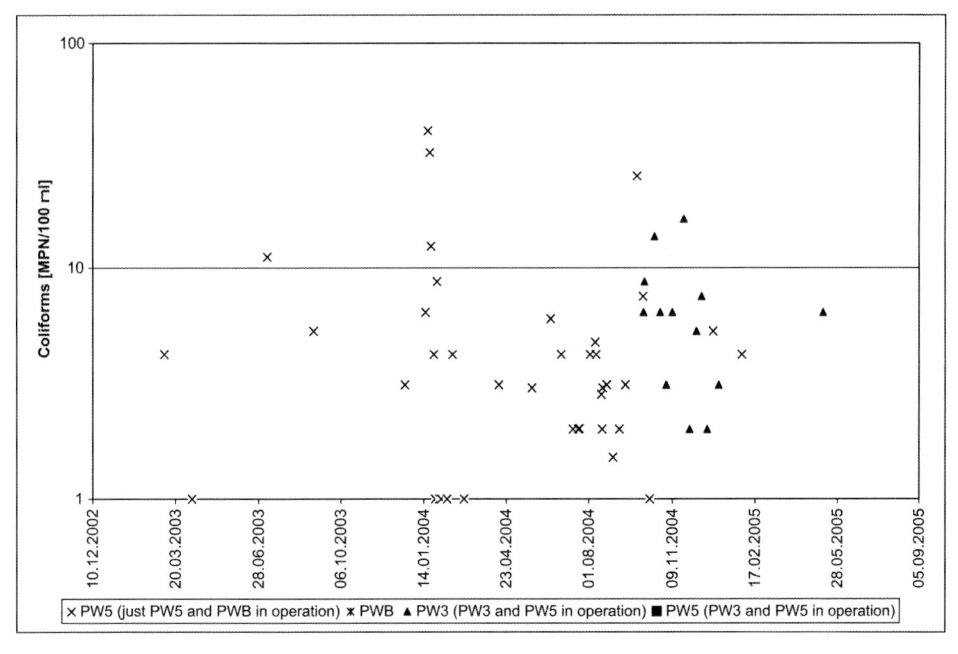

Figure 11: *Occurrence of coliforms in the different pumping stations depending on the operation mode.*

2.4 Investigation of Caisson wells

The Caisson wells are masonary: i.e. built from stone on stone and not plastered to ensure water exchange with the surrounding soil and rock matrix of the well. In March 2006, and August September 2006 the Caisson wells were examined by taking bailed samples in periods of normaloperation and when not operating. In addition to the water level of the wells, temperature, conductivity and oxygen of the water were measured.

The coliform contamination in the different wells varied (Figure 13) without correlation with: the water level in the wells; their position to the River Rhine; the pumping station; and/or to each other (Figure 12). Additionally, there was no difference in their coliform contamination, independent of whether the wells were in operation or not. Between 28.08.2006 and 01.10.2006 the wells were inoperative. Even after 4 weeks of stagnation there was no difference in the contamination characteristics.

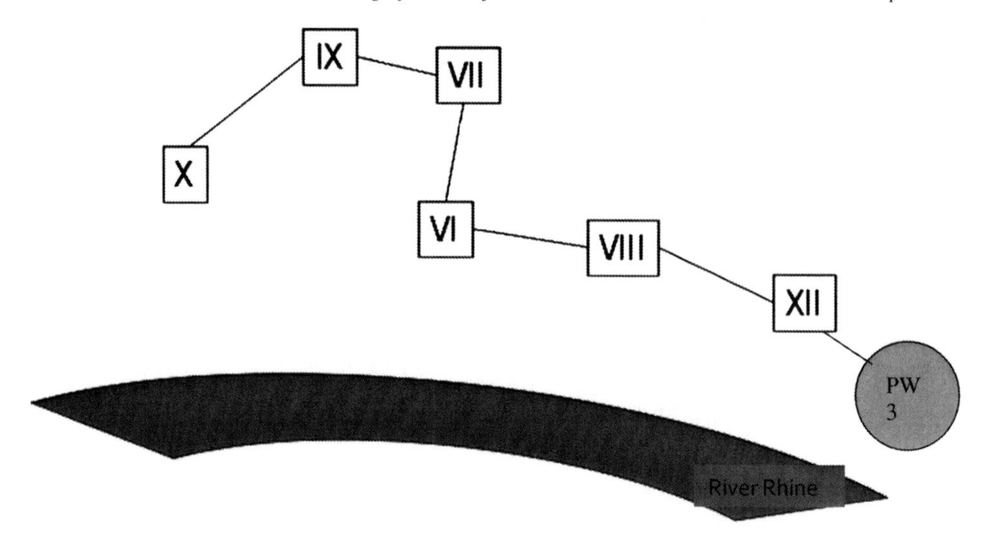

Figure 12: *Floor plan of Caisson wells.*

E.coli was detected once (well 10, 29.08.2006) when, because of a small flood wave, the water level was higher during the sampling in March than at the second testing in August/September.

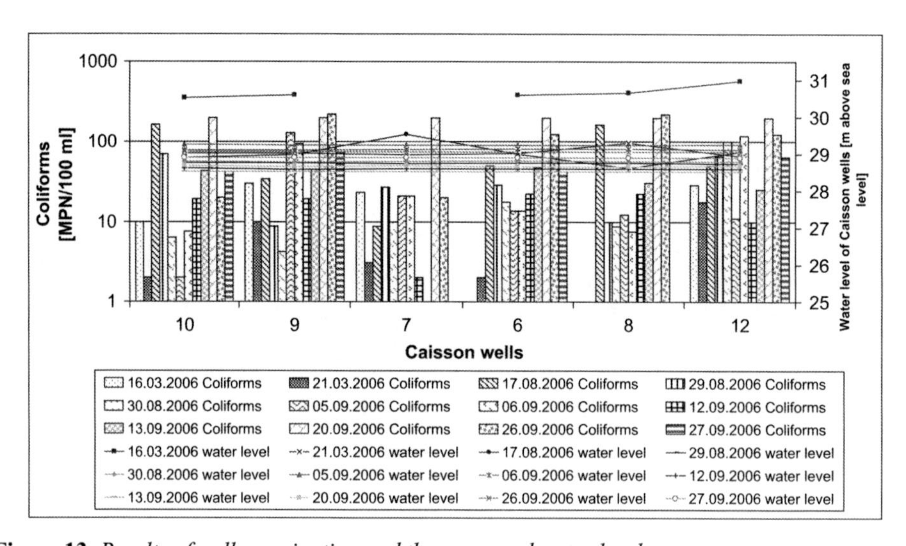

Figure 13: *Results of well examination and the measured water level.*

An obvious influence of the water level of the River Rhine on the water level of the wells was indentified (Figure 14).

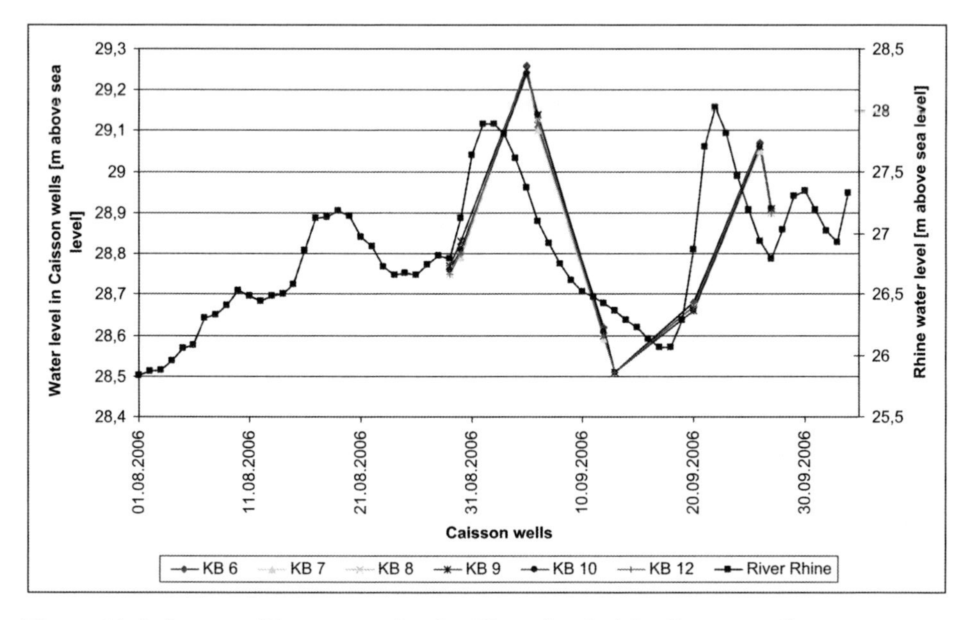

Figure 14: *Influence of River water level to Water level of the Caisson wells.*

2.4.1 Influence of rainfall on the coliform contamination of Caisson wells

Because of the cut-and-cover method of the Caisson wells, the influence of rainfall on the coliform contamination was assessed. An influence could not be detected (Figure 15). Even in periods with no rainfall coliforms occurred in the wells. And heavy rainfalls have not led to an increasing coliform contamination.

2.5 Influence of Caisson wells on coliform contamination of raw water

During the evaluation period, the contamination of the raw water was observed four days a week. In the period of abandonment of the Caisson wells (28.08.2006 – 01.10.2006) less coliforms were detected in the raw water of the pumping stations than during the operation time of the wells (Figure 16).

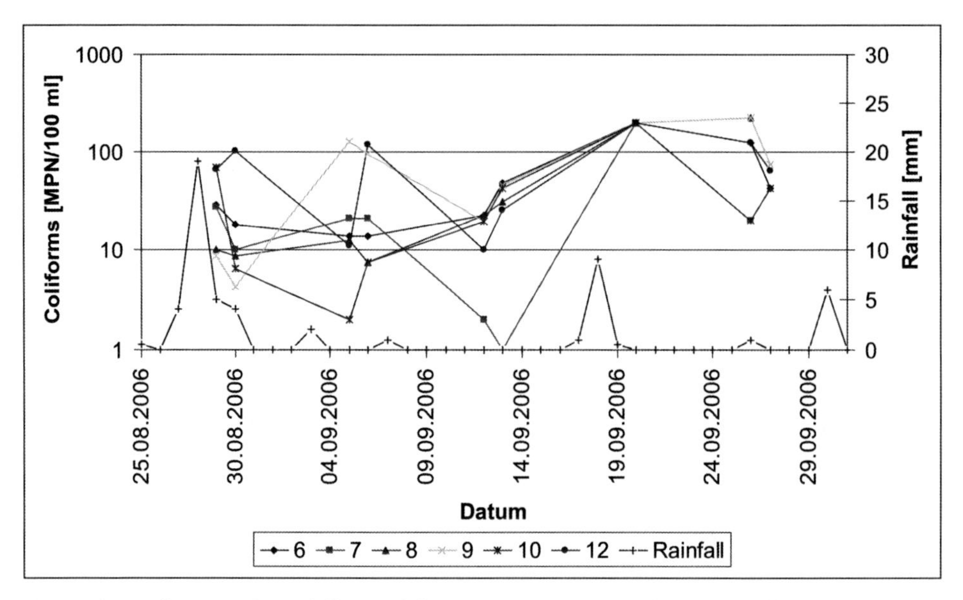

Figure 15: *Influence of rainfall on coliform contamination in Caisson wells.*

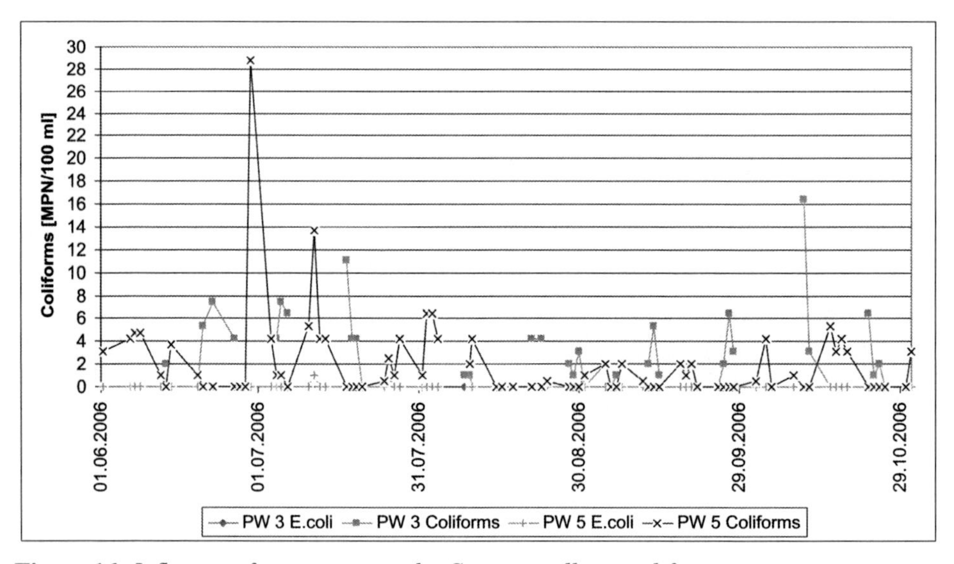

Figure 16: *Influence of not operating the Caisson wells on coliform contamination in raw water of the pumping stations.*

The coliforms still occurred just in the raw water of pumping station 5 when pumping station 3 was not operated. A damaged valve caused a slide discharge. After changing the valve the Caisson wells were operated for nearly two months and put out of operation again on 27.11.2006 (Figure 17). Even in times of altering water level in the River Rhine the coliform contamination decreased in the raw water.

The demonstrated influence of the inoperative Caisson wells on the coliform contamination of the pumping stations was also evident in the raw water of the treatment plant (Figure 18).

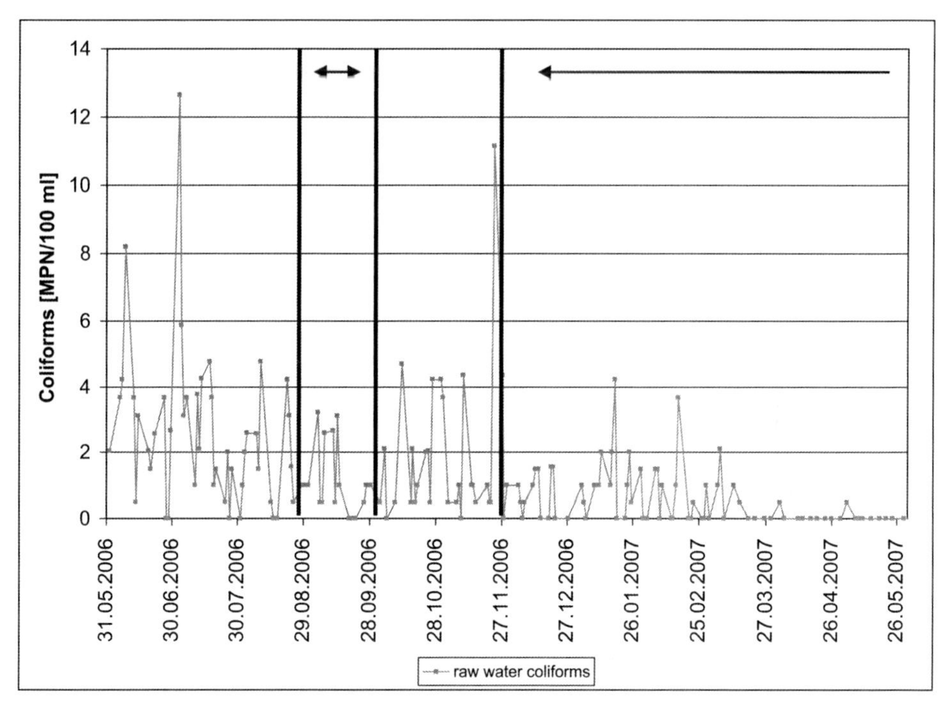

Figure 17: *Influence of operating and not operating the Caisson wells on coliform contamination in raw water of the treatment plant (the arrows mark the inoperative periods).*

Since March 2007 just a few coliforms appeared in the raw water of the treatment plant (Figure 17) and perhaps a longer period of abandonment of the Caisson wells was necessary to decrease the coliform contamination in the raw water. On the other hand, both - the water level of the River Rhine and its number of coliforms - was constantly low in this period.

The coliforms that appeared were detected at times of changing water level of the River Rhine. To approve the results that the Caisson wells were one source of the contamination of the raw water on 06.08.2007 the wells were put into operation. In this period strongly changing water level of the River Rhine was not expected so that the results should demonstrate just the influence of the wells.

After operation both - in the raw water of the pumping stations and in raw water of the treatment plant - the number of coliforms increased (Figure 18). But this slope was not as high as the one in 2006 after putting the wells into operation. With the decrease of water level the number of coliforms decreased in the raw water of the pumping stations and the treatment plant.

On 18.09.2007 the pipeline between the pumping stations 5 and 3 was interrupted to change a valve. Thereby pumping station 3 and both the Caisson wells and the shaft wells to number 75 were separated from operation. Since then no coliforms appeared in the raw water (Figure 18).

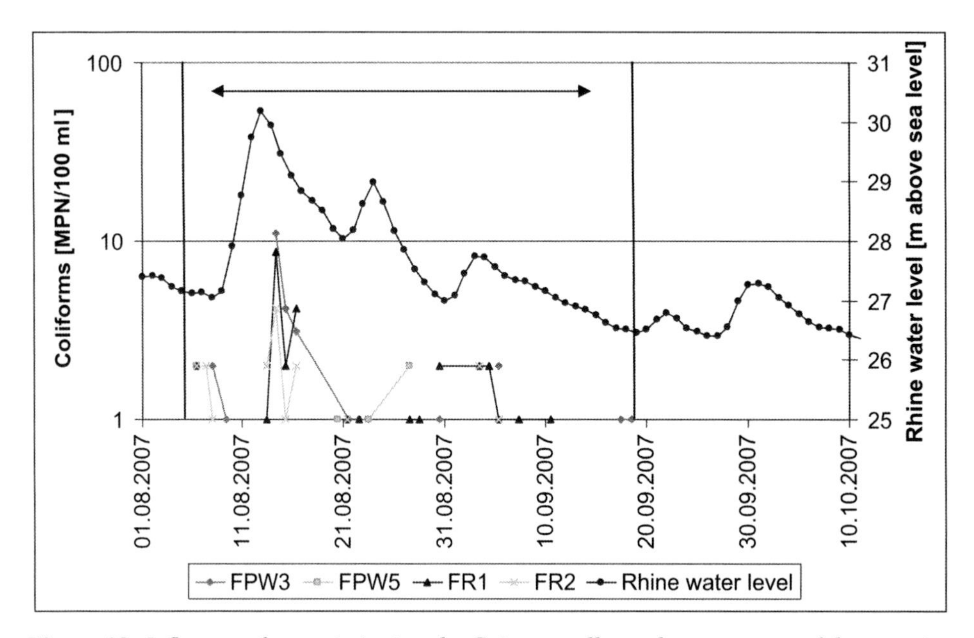

Figure 18: *Influence of commissioning the Caisson wells on the raw water of the pumping stations and treatment plant (the arrow marks the period of operation).*

2.6 Survey of shaft wells

The results identified the Caisson wells as a source for the coliform contamination of the raw water. But even when these wells were not operated, coliforms were detected. The existence of another source of contamination could be assumed. Therefore the shaft wells which are located near to the River Rhine in the catchment area of pumping station 3 were examined more intensively (Figure 10).

In August 2006 bailed samples of wells 31 to 38 were taken. But no coliforms could be detected. The reason was that the samples were not taken from the filter drain, but from a pipe nearby with stagnating water of indefinite age. Therefore the absence of coliforms was no surprise. In a second survey, samples were taken from the filter drain of wells 31 and 32 using a pump. But no coliforms were detected. These two samples did not suffice to detremine the possible influence of the vertical wells. Further examination was declined because of its complexity. In order to take the samples the well gallery had to be aerated, entering itself needed two persons to take the samples and one to ensure safe access to the well gallery which was not possible during flood periods. Additionally just the wells 31 and 32 had the possibility to take samples from the filter drain and back fitting the other wells would have been too expensive.

To verify and confirm the influence of the shaft wells on the coliform contamination of raw water, a special mode of operation was established. During November and December

2007 the Caisson and shaft wells and also pumping station 5 was not operated so that pumping station 3 derived its raw water just from shaft wells 41 to 74. During this investigation, pumping station 3 and the raw water of the treatment plant was sampled daily Monday to Thursday and analysed (Colilert®).

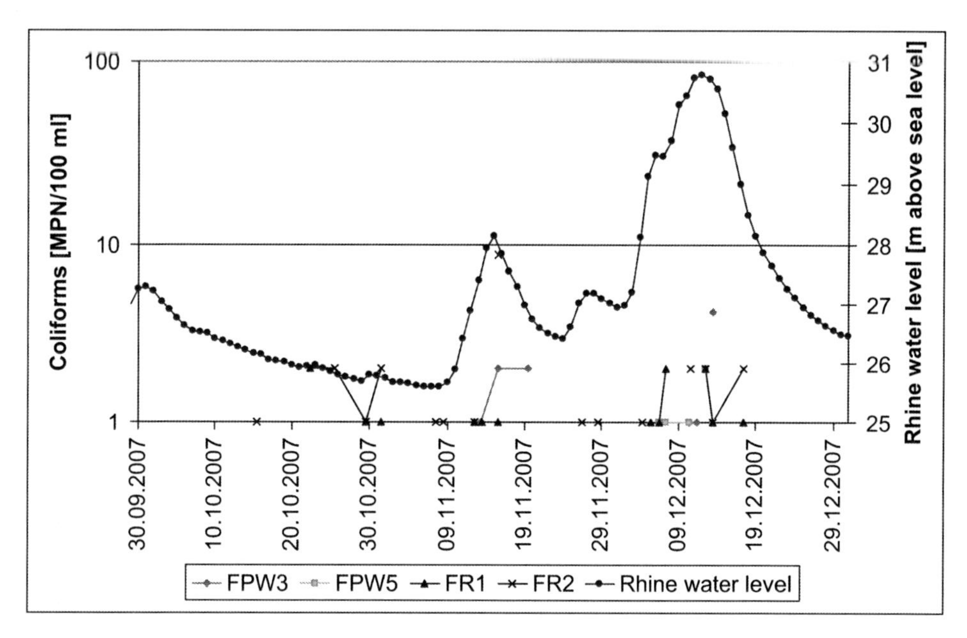

Figure 19: *Number of coliforms in the raw water of both pumping stations and water treatment plant.*

If pumping station 3 was operated just with the shaft wells at the furthest distance to the River Rhine in periods of rapidly increasing water levels, coliforms could be detected in the raw water of the pumping station (Figure 19). During 04.12.2007 and 13.12.2007 pumping station 5 was put into operation and coliforms were detected in raw water. It was not possible to evaluate whether the activation of pumping station 5 or the increasing water level of the River Rhine was the reason for contamination.

2.7 Evidence of *E.coli*

During the whole survey including the data search, *E.coli* was detected rarely and in low numbers (Figure 20). The appearance of *E.coli* in the raw water of the pumping stations normally indicated their occurrence in the raw water of the treatment plant (Figures 20 and 21). But there was no correlation between frequency and number of appearance.

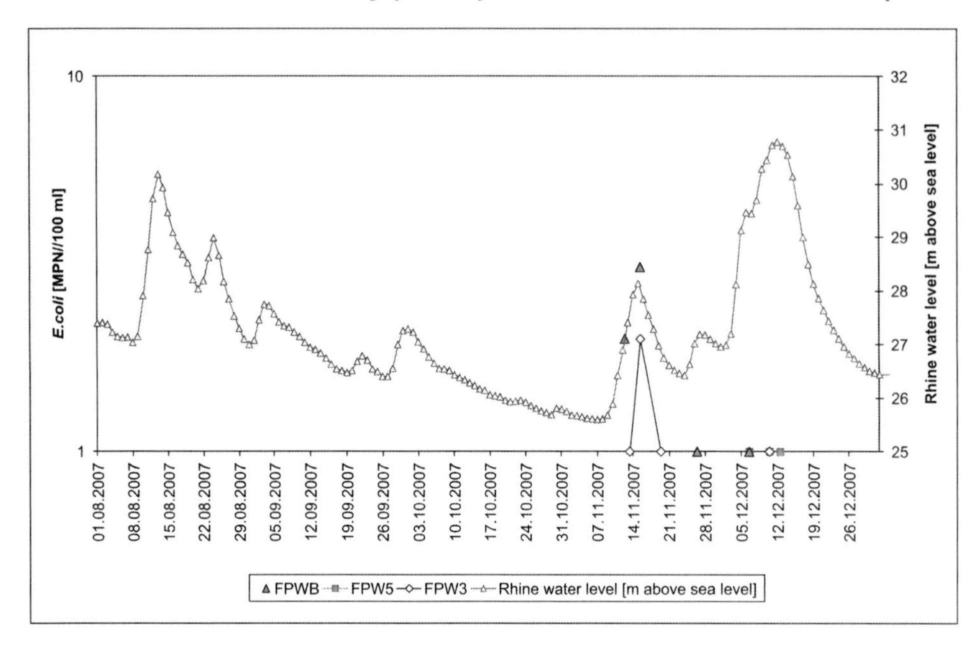

Figure 20: *Occurrence of E.coli in raw water of the pumping stations.*

Normally *E.coli* was detected in periods of increasing or decreasing Rhine water level. Therefore *E.coli* seemed to be an index parameter for a strong influence of the River Rhine on the raw water quality (Figures 20 and 21).

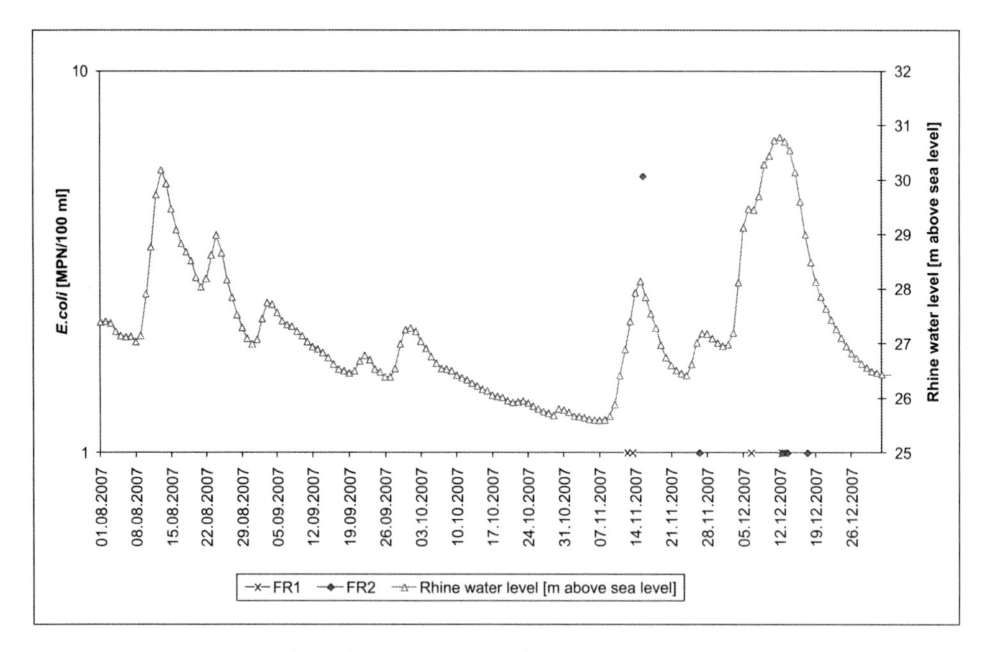

Figure 21: *Occurrence of E.coli in raw water of the water treatment plant.*

2.8 Evidence of other faecal indicators

Besides the coliforms additional faecal indicators were determined.

2.8.1 Intestinal Enterococci

In raw water intestinal Enterococci were detected seven times at values of of 1 enterococcus/100 ml since 2004. There was no correlation with the appearance of coliforms.

2.8.2 Coliphages

Coliphages were analysed in river water and in the raw water of the water treatment plant. In periods with low water levels in the River Rhine no coliphages were detected in the raw water (Figure 22).

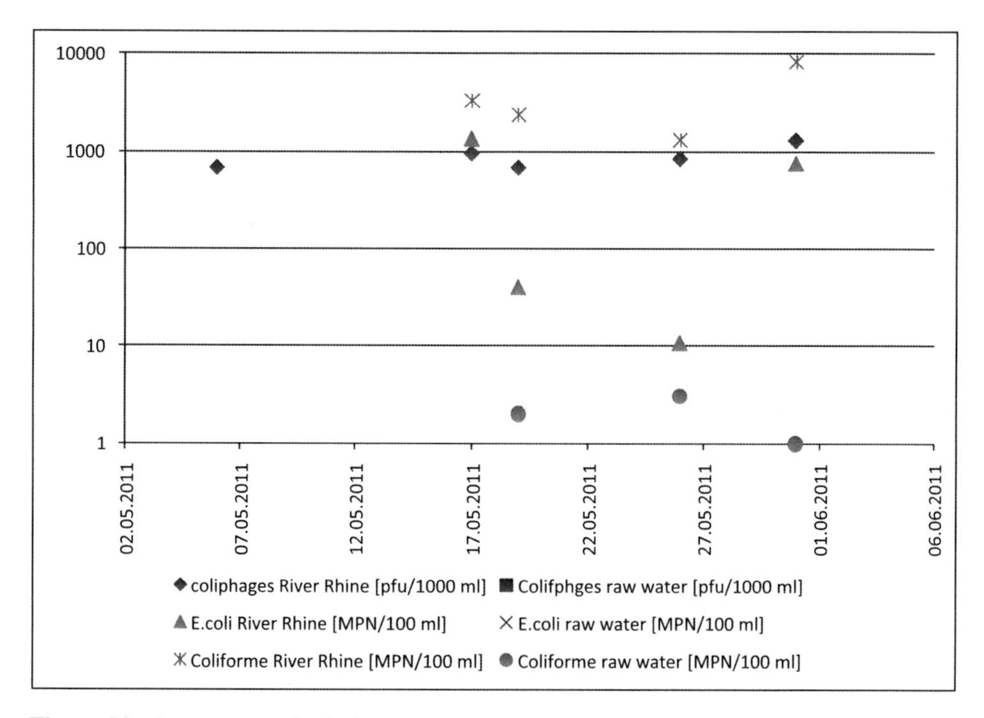

Figure 22: *Appearance of coliphages and coliforms in River water and raw water.*

3 CONCLUSION

The different methods for coliform analysis led to results with different concentrations. But even if the occurrence of coliforms depended on the analytical method, a contamination of the raw water of the water works existed.

The risk assessment was complex. Sometimes the mode of operation could not be changed as scheduled, because of operational interest and often it was not possible to eliminate special influences on their assessment. Different parts of the water catchment of the water

works at Flehe had an influence on the raw water quality. The masoned and not plastered Caisson wells had sometimes a high coliform contamination and were highly influenced by the Rhine water level. Because no adequate facilities to sample the shaft wells were available, their contamination could not be assessed. But their influence on the raw water quality was determined indirectly by changing modes of operation. One surprising result was that not only the shaft wells near by the River Rhine were an origin of the coliforms in raw water. Even the shaft wells at a greater distance from the river influenced the raw water quality. Not operating several parts of the water catchment led to lower numbers or less evidence of coliform contamination of the raw water but was no guarantee for coliform free water. Depending on the mode of operation, the location of the appearance of coliforms in raw water could be determined. For example when both pumping stations 3 and 5 were operated, coliforms only appeared in the raw water of pumping station 3.
Rapidly changing water levels of the River Rhine always led to a coliform contamination of the raw water regardless of the mode of operation. Therefore, the water level of the River Rhine has to be monitored as well as the coliform contamination of raw water. The measurement of enterococci was unsuitable for the risk assessment because of they were only rarely detected. Coliphages seemed to be inappropriate as well, because they could not be detected even if coliforms appeared.
The river bank filtration was an effective treatment step for eliminating coliforms. The passage through the bank reduced the coliform contamination of the River Rhine to less than 100 coliforms/ 100 ml in case of the coliforms and to under 10 cfu/100 ml in case of *E.coli*.
In order to produce safe drinking water, a disinfection step during the water treatment is necessary in the water works at Flehe.

References

1 Bundesgesundheitsbl 2009 52:**474 – 482** DOI 10.1007/s00103-009-0823-7
2 D. Schoenen, N. Eckschlag and G. Pankroff, GWF 149 (2008) Nr. 11, **858 - 863**

MICROBIOLOGY OF SUSTAINABLE WATER SYSTEMS; RAINWATER HARVESTING – A UK PERSPECTIVE

L. Fewtrell[1], C. Davies[2], C. Francis[2], H. Jones[3], J. Watkins[2] and D. Kay[1]

[1]Aberystwyth University
[2]CREH Analytical Ltd, Leeds UK
[3]MWH Ltd, Cardiff, UK

1 INTRODUCTION

Rainwater harvesting systems have been deployed in urban centres world-wide, at a range of scales, to provide a water resource for non-potable domestic and industrial uses, such as garden watering, open-space irrigation, and toilet flushing. An additional benefit of such water harvesting systems is that they can reduce water flux within surface water drains and, where 'combined' sewage systems are installed, they reduce the flow in these systems by the storm flow volume retained in the collection system. This 'flow reduction' during storm events is potentially beneficial in reducing urban flooding from storm sewer discharges and related infrastructure damage caused by excess hydrological loadings.

While harvested rainwater is not generally used as drinking water in the UK, replacement of 'potable quality' treated water in domestic settings with untreated 'environmental' waters raises questions of water quality and any potential infection risks incurred by residents of properties in which harvested rainwater is used for non-potable uses.

This chapter reviews the literature data on the microbiology of rainwater in developed countries, before outlining recent empirical data acquisition exercises and presenting an exploratory quantitative microbial risk assessment (QMRA) to assess the possible health impact of using harvested supplies.

2 LITERATURE REVIEW

The main sources of external contamination of rainwater supplies are pollution from the air, bird and animal droppings and insects. In the UK, the principal source of microbial contamination is likely to be avian. A number of studies have shown that roof-water quality varies, with the greatest pollution load being related to the first flush during rainstorms.[1] Faecally contaminated debris can build up on roofs during dry periods, meaning that the initial runoff is likely to be full of debris, sediment and highly turbid. Rainfall intensity and the number of dry days preceding a rainfall event can markedly affect the quality of runoff water. At a single property in Australia, the faecal coliform levels measured in first flush events varied widely over 11 rainfall events.[2] Discarding the first 1 mm of runoff reduced the bacterial load entering the tank by between 9 and 62% for individual rainfall events. Stored water, however, still contained significant levels of contamination, with faecal coliform concentrations up to 480 cfu/100ml. The results from studies measuring faecal indicator organisms in roof-water and stored rainwater in developed countries are summarised in Table 1.

Table 1 *Presence of faecal indicator bacteria in roof-water and stored rainwater in developed countries.*

Setting	Parameter	Sample location	(/100ml)	Comments
Europe				
England, urban area[3]	*E. coli*	run-off	Prev 94% (n=88) Range 0 – 16000 Median 52	Rainwater collected (prior to treatment) from the roof of the Millennium Dome
	Enterococci	run-off	Prev 97% (n=34) Range 0 – 680 Median >200	
England, semi rural[4]	*E. coli*	stored water	Prev 35% (n=26) Range 0-53	Samples taken from outside tap, possibility of mains water top-up
	Faecal streptococci	stored water	Prev 58% (n=26) Range 0 - 79	
Scotland, rural[5]	Thermotolerant coliforms	run-off	GM mean range 383-3627 (n=68)	Sampling from farm roofs
	Faecal streptococci	run-off	GM mean range 482-46580 (n=68)	
Denmark, urban[6]	E. coli	stored water	Prev 79% (n=14) Range 4 – 990 Median 245	Samples taken from 7 rainwater harvesting systems
Germany, urban[7]	E. coli	stored water	n=972 Median 26 Max >10000	102 rainwater tanks
	Faecal streptococci	stored water	n=969 Max >10000	
Australasia				
Australia, rural[8]	E. coli	stored water	Prev 26% (n=47) Range 0 – 370 Mean 15	
Australia, urban[9]	Thermotolerant coliforms	stored water	n=30 Max 800 Mean 119	
Australia[10]	Thermotolerant coliforms	run-off	Prev 100% (n=24) Range 75-125	Data presented in graphical form
Australia[11]	E. coli	stored water	not detected	
New Zealand, rural[12]	E. coli	stored water	Prev 38% (n=24) Range 0-111	
New Zealand, rural[13]	Thermotolerant coliforms	stored water	Prev 56% (systems) Range 0-840	125 systems, sampling from household tap
	Enterococci	stored water	Range 0 – 4900	
USA				
USA, rural[14]	Thermotolerant coliforms	stored water	Prev 3% (tanks) Typical values 10-20	30 tanks sampled

In addition to faecal indicator organism monitoring a number of studies also looked at the prevalence of selected pathogens in roof-water and stored rainwater, as shown in Table 2 (adapted from Sinclair et al 2005 and Fewtrell and Kay 2007).[15,16]

Table 2 *Pathogen prevalence in rainwater samples.*

Setting	Sample location	Parameter	Prevalence
Europe			
England, urban[3]	run-off	*Campylobacter* spp. *Cryptosporidium* spp. *E. coli* O157 *Giardia* spp. *Salmonella* spp. *Shigella* spp. *L. pneumophila*	0% (n=2) 0% (n=2) 0% (n=2) 50% (n=2) 0% (n=2) 0% (n=2) 0% (n=2)
Denmark, urban[6]	stored water	*Campylobacter* spp. *Cryptosporidium* spp. *Giardia* spp. *Aeromonas* spp. *Legionella* spp. *L. pneumophila* *M. avium* *P. aeruginosa*	12% (n=17) 35% (n=17) 0% (n=17) 14% (n=14) 71% (n=7) 0% (n=14) 7% (n=14) 7% (n=14)
Germany, urban[7]	stored water	*Campylobacter* spp. *Salmonella* spp. *Shigella* spp. *P. aeruginosa* *Legionella* spp.	0% (n=142) 0.1% (n=798) 0% (n=342) 11% (n=710) 0% (n=418)
Australasia			
Australia, rural[8]	stored water	*Campylobacter* spp. *Cryptosporidium* spp. *Giardia* spp.	13% (n=42) 0% (n=20) 0% (n=20)
Australia, urban[9]	stored water	*Pseudomonas* spp.	100% (n=18)
Australia[11]	stored water	*Salmonella* spp. *Shigella* spp.	0% (n>60) 0% (n>60)
Australia[17]	stored water	*Mycobacterium* spp.	35% (n=205)
New Zealand, rural[12]	stored water	*Campylobacter* spp.	38% (n=24)
New Zealand, rural[13]	stored water	*Campylobacter* spp. *Cryptosporidium* spp. *Giardia* spp. *Salmonella* spp. *Aeromonas* spp. *Legionella* spp.	0% (n=115) 4% (n=50) 0% (n=50) 0.9% (n=115) 16% (n=125) 0% (n=23)

Analyses were conducted for a total of ten different pathogens. *Campylobacter* was the pathogen most frequently sought (6/9 studies), followed by *Salmonella* spp., *Legionella* spp., *Cryptosporidium* spp. and *Giardia* spp. (4/9 studies each). The opportunistic pathogens, *Pseudomonas* spp., *Mycobacterium* spp. and *Aeromonas* spp., were found in some samples in every study that chose to look for them.

3 RAINWATER QUALITY STUDY

3.1 Study Design

Samples were taken from three wall-mounted rainwater harvesting systems (Figure 1) over a nine week period from mid-April to mid-June 2009, one in Leeds (50 sample runs, see Figure 1) and two in Wales (6 sample runs). The tanks had a volume of 100 litres and were manufactured by Aqualogic © specifically for rainwater harvesting applications. The purpose of the tank is to buffer storage of rainfall events to ensure availability of untreated water for toilet flushing during dry periods. Clearly, however, this pattern of use implies the potential for microbial die-off and protection within biofilms formed within the tank environment. Thus the utility of this study is in the empirical data reported which is more accurately described as stored roof drainage quality rather than immediate runoff quality from roof systems during rainfall events.

Figure 1 *Installation of the rainwater harvesting tank at the Leeds (UK) laboratory and the meteorological station used to record rainfall and other parameters.*

At all three sites, harvested rainwater stored in the tanks was analysed for the following microbial parameters, using standard methods,[18-21] and turbidity (NTU):

- Total coliforms
- *Escherichia coli*
- Enterococci
- *Clostridium perfringens*
- *Pseudomonas aeruginosa*

- Plate count at 2 days (37°C)
- Plate count at 3 days (22°C)
- *Salmonella* spp.
- *Campylobacter* spp.

3.2 Results

The results of the faecal indicator organism (FIO) analysis are shown in Table 3.

Table 3 *Summary of faecal indicator organism concentrations.*

FIO/site	GM[a]	Min	Max	n[b]	Prev[c] (%)
Coliforms (/100ml)					
Leeds	7	0	14 000	50	58
Nelson	21	0	630	6	83
Cilfynydd	12	0	650	6	50
All samples	8	0	14 000	62	60
E. coli (/100ml)					
Leeds	6	0	14 000	50	58
Nelson	18	0	630	6	67
Cilfynydd	14	0	650	6	50
All samples	7	0	14 000	62	58
Enterococci (/100ml)					
Leeds	18	0	1700	50	80
Nelson	53	0	1000	6	83
Cilfynydd	16	0	230	6	83
All samples	20	0	1700	62	62
C. perfringens (/100ml)					
Leeds	7	0	50	50	96
Nelson	0.3	0	3	6	16
Cilfynydd	14	2	45	6	100
All samples	6	0	50	62	89
2 day plate count (/ml)					
Leeds	190	10	3600	50	100
Nelson	470	190	2700	6	100
Cilfynydd	190	40	480	6	100
All samples	210	10	3600	62	100
3 day plate count (/ml)					
Leeds	19 000	1000	170 000	50	100
Nelson	31 000	14 000	250 000	6	100
Cilfynydd	29 000	8800	160 000	6	100
All samples	21 000	1000	250 000	62	100

[a] Geometric mean, 10^X, where X is the mean value of the $\log_{10}$ transformed values
[b] Number of observations
[c] Prevalence (i.e. percentage of positive observations)

Salmonella spp. were not present in any of the samples (n=27). *P. aeruginosa* was isolated from a single sample (n=62), with a concentration of 4/100ml. *Campylobacter* spp. were detected in 16 samples (n=35). Over 50% of the samples analysed (n=23) for *Campylobacter* spp. from the Leeds site were positive for the pathogen, with concentrations ranging from 1 to 91/litre. Concentration values for the positive *Campylobacter* spp. samples at Nelson and Cilfynydd were not available as only presence/absence was determined. All of the *Campylobacter* spp. positive samples occurred after the middle of May.

The rainwater harvesting system at Leeds was sampled more frequently than the other two sites, so it was possible to group the FIO results based on the rainfall (mm) recorded at the site during the 24 hours prior to the sampling (Table 4). The cut-point for the comparison was 5mm, with < 5mm indicating 'dry weather' and ≥ 5mm deemed to be 'wet weather'. It can be seen from this Table (and Figure 2, which shows the geometric mean values) that coliforms, *E. coli* and enterococci were all significantly elevated during the wet weather sampling.

Table 4 *Summary of faecal indicator organisms by pre-sample rainfall.*

FIO/rainfall category	GM[a]	SD[b]	Min	Max	n[c]
Coliforms (/100ml)					
dry	4	1.0693	0	14 000	42
wet	82*	1.0154	3	1400	8
E. coli (/100ml)					
dry	4	1.0676	0	14 000	42
wet	43*	1.0322	1	1200	8
Enterococci (/100ml)					
dry	11	0.8053	0	380	42
wet	220*	0.6192	26	1700	8
C. perfringens (/100ml)					
dry	6	0.3924	0	50	42
wet	9	0.4082	1	31	8
2 day plate count (/ml)					
dry	180	0.7404	10	3600	38
wet	250	0.4022	80	1600	8
3 day plate count (/ml)					
dry	21 000	0.5395	1000	170 000	42
wet	10 000	0.3509	2600	27 000	8

a Geometric mean, 10^{X}, where X is the mean value of the $\log_{10}$ transformed values
b Standard deviation of $\log_{10}$ transformed values
c Number of observations
* Geometric mean significantly elevated compared to dry weather geometric mean (Student's t-test, $p < 0.05$)

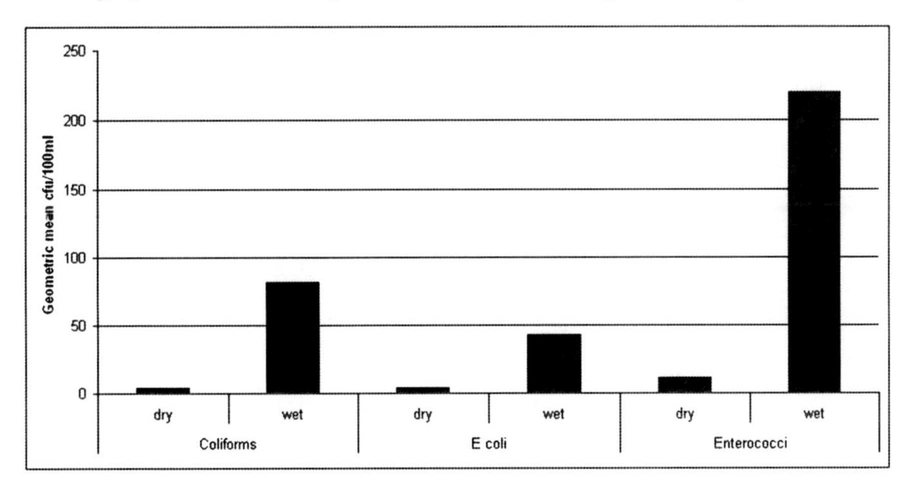

Figure 2 *Geometric mean 'wet' and 'dry' values for coliforms, E. coli and enterococci at the Leeds site.*

4 QMRA

An exploratory QMRA was performed using the rainwater harvesting data collected from the Leeds site, assessing the risk to health (in a hypothetical population) from flushing toilets with harvested rainwater potentially contaminated with *Campylobacter* spp..

4.1 Exposure Scenario

The population assumed to be exposed to the rainwater harvesting was loosely based on the population of a newly built development in the south of England, comprising approximately 1900 houses.[22] Thus, for illustrative purposes in this analysis, we have sought to simulate a moderate-scale implementation of the technology within a new development and/or retro-fitting within an existing housing development. We have assumed 100% adoption within the development.

4.2 Hazard Characterization (reference pathogen) and Dose-response Relationship

As mentioned earlier, the principal source of pathogens in harvested supplies in the UK is likely to be avian. The intestinal carriage of a number of pathogens by wild birds has been recognised for a long time.[23] The two most commonly studied and (where a variety of pathogens have been examined) most frequently isolated from birds are *Salmonella* spp. and *Campylobacter* spp.[24-26] In the studies summarised in Table 3, *Campylobacter* spp. were isolated with prevalences ranging between 0 to 38%, compared to 0 to 0.9% for *Salmonella* spp. Also, *Salmonella* spp. were not isolated from any of the samples in the current study, thus *Campylobacter* spp. was chosen as the hazard to investigate in this QMRA.

Campylobacter is the most common cause of bacterial gastroenteritis in England and Wales (Figure 3) and the illness, campylobacteriosis, is characterized by severe diarrhoea and abdominal pain. In some cases, secondary adverse health outcomes, such as Guillain-Barré syndrome (inflammation of the nerves, which may result in paralysis), may occur as a result of the infection.[27]

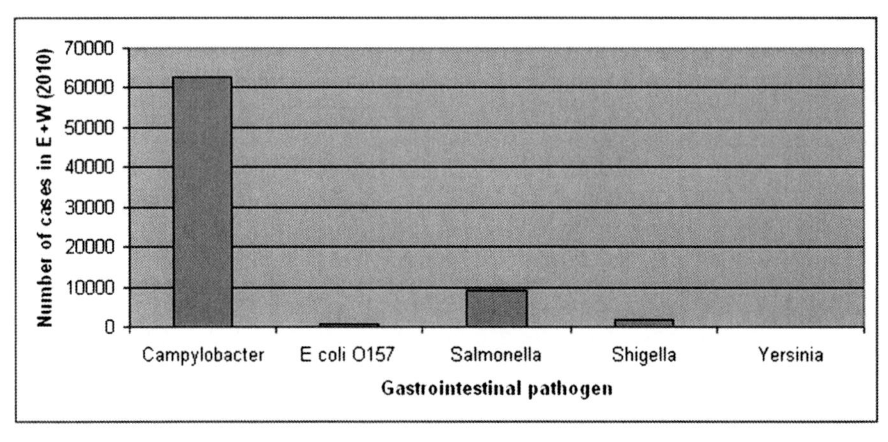

Figure 3 *Laboratory reports of bacterial gastrointestinal illness by pathogen in 2010.*

Dose-response functions describe the quantitative relationship between dose and outcome (i.e. infection). A dose-response (β-poisson) model for *Campylobacter* spp. has been developed based on experimental data from healthy volunteers.[28] Infection precedes illness and, based on the literature, it has been assumed that 30% of infections result in illness.[29]

4.3 Exposure Assessment

Two different exposure scenarios, based on different *Campylobacter* spp. concentrations and frequency of contamination, were explored.

- A 'worst case' scenario, based on a *Campylobacter* spp. concentration of 91/litre (the highest concentration recorded at the Leeds site) and a frequency contamination of 50%.
- An alternative scenario based on the median of the positive Campylobacter values (12/litre), and an assumed contamination frequency of 33%. This contamination frequency is more typical of the high end values seen in the published literature (0 – 38%)[22] and is the value recorded from the two Welsh sites.

Flushing a toilet produces an aerosol, which may spread microorganisms present in the toilet bowl around the surrounding area. It has been demonstrated in seeding experiments that microorganisms can be ejected during flushing, to a height of at least 83 cm above the seat,[30] a height that could result in pathogens being in a position where they could be ingested. The volume of water ejected during a 'typical' flush is unknown, but is likely to be small and probably between 1 and 2ml. Only a proportion of this is likely to reach a susceptible host; 10% has been assumed. It has been assumed that people are exposed to flush aerosol 5% of the time (i.e. 1 flush in 20). The frequency of toilet use is assumed to be between 3 and 6 times a day, with the range accounting for home workers and those who work away from home.[31] Children under the age of 3 are not thought to be exposed as they are unlikely to be toilet trained.[32]

4.4 Risk Characterization

This brings together the hazard characterization, dose-response assessment and exposure assessment to produce an estimate of the probability of infection and from that the actual expected clinical illness in the exposed population over the period of a year.

- The worst case scenario results in an estimate of nine cases of campylobacteriosis (in a population of almost 5000) on an annual basis.
- The alternative scenario suggests an estimate of less than 1 case a year in the test population.

These values can be contextualised by considering the annual number of cases of campylobacteriosis in England and Wales on a population basis. While Figure 3 suggests that the annual number of cases is just over 60,000 this figure is only based on laboratory reports and thus is a major underestimate of the actual number of cases experienced by the population. Adak et al 2002, took the results of the Infectious Intestinal Disease study, which was conducted in England in 2000, and determined that the ascertainment ratio for campylobacteriosis was 10.3, thus the true number of cases of campylobacteriosis in England and Wales in 2010 was closer to 645,000.[33] Using an approximate population estimate of 54 million, this equates to almost 12 cases/1000 population. The worst case scenario is equivalent to approximately 2 cases/1000 in the case study population, while the more realistic scenario is equivalent to approximately 0.18/1000 in the case study population.

5 CONCLUSIONS

There is relatively little published data on the quality of harvested rainwater supplies, although the available evidence does suggest potential for contamination with both faecal indicator organisms and pathogens.

This empirical study found that all sites examined experienced periods of elevated faecal indicator concentrations which would cause regulatory concern if, for example, the public were exposed to these concentrations in a regulated 'environmental' bathing water.[34] None of the three sites would be compliant with microbial standards required of a treated drinking water.[29] While unsurprising for a raw environmental water, this finding is worth stating given the common perception that harvested rainwater is 'green' and 'clean', and thus is 'safe' in sanitary terms.

It is also worth noting the relatively high values observed on the 3 day plate count incubated at 22°C. This parameter is often used to indicate the potential for microbial re-growth within distribution systems which could be of relevance to raw waters stored in the harvesting systems. This aspect may be worthy of further examination.

The QMRA, based on a median *Campylobacter* concentration of 12/litre, demonstrated only a virtually negligible risk from using harvested rainwater supplies for toilet flushing, the same was not true, however, for the worst case scenario. Although no formal sensitivity analysis was conducted, the pathogen concentration is the key factor in the estimate of health impacts. Although this study presents some pathogen data, water quality is still poorly characterised and it would be useful to have pathogen data over a period of at least a year (to examine seasonality) and also at several different locations.

Given the occasional high concentrations of *Campylobacter* spp. isolated from the rainwater at the Leeds site it is strongly recommended that harvested supplies are not used as sources of drinking water without ensuring adequate treatment.

Acknowledgements

We are grateful to Aqualogic who provided the rainwater harvest tank technology and the Dwr Cymru Welsh Water who funded this investigation

References

1 J. Förster, *Water Sci. Technol.*, 1999, **39**, 137.
2 T. Gardner, J. Baisden and G. Millar, in *Sustainable Water in the Urban Environment, 2004 Conference*, Brisbane, Australia, 2004.
3 R. Birks, J. Colbourne, S. Hills and R. Hobson, *Water Sci. Technol.*, 2004, **50**, 165.
4 M.M. Day, *A report on 'Freerain', the rainwater recycling system*, report for Gusto Construction Ltd., Severn Trent Water and the Environment Agency, 2002.
5 D. Kay and A.C Edwards, *Evaluation of bacterial loads from farmyard drainage systems, 2003*, report to the Scottish Executive, Centre for Research into Environment and Health, Wales, 2003.
6 H.J. Albrechtsen, *Water Sci. Technol.*, 2002, **46**, 311.
7 R. Holländer, M. Bullermann, C. Groß, H. Hartung, K. König, F.-K. Lücke and E. Nolde, *Gesundheitswesen*, 1996, **58**, 288.
8 R. Bannister, J. Westwood and A. McNeill, *Water ECO science report no. 139/97*, 1997.
9 P.J. Coombes, G. Kuczera and J.D. Kalma, in *Proceedings of the 3rd International Hydrological and Water Resources Symposium*, Perth, Australia, 2002.
10 P.R. Thomas and G.R. Green, *Water Sci. Technol.*, 1993, **28**, 291.
11 R. Thurman, *Australian Microbiol.*, 1995.
12 M.G. Savill, J.A. Hudson, A.A. Ball, J.D. Klena, P. Scholes, R.J. Whyte, R.E. McCormick and D. Jankovic, *J. Appl. Microbio.*, 2001, **91**, 38.
13 G. Simmons, V. Hope, G. Lewis, J. Whitmore and W. Gao, *Water Res.*, 2001, **35**, 1518.
14 D.J. Lye, *Water Res. Bull.*, 1987, **23**, 1063.
15 M.I. Sinclair, K. Leder and H. Chapman, *CC for Water Quality and Treatment; Occasional Paper 10*, 2005.
16 L. Fewtrell and D. Kay, *Urban Water Journal*, 2007, **4**, 253.
17 R.E. Tuffley and J.D. Holbeche, *Appl. Environ. Microbiol.*, 1980, **39**, 48.
18 SCA, *Methods for the Examination of Waters and Associated Materials*, Environment Agency, 2009.
19 SCA, *Methods for the Examination of Waters and Associated Materials*, Environment Agency, 2007.
20 SCA, *Methods for the Examination of Waters and Associated Materials*, Environment Agency, 2006.

21 SCA, *Methods for the Examination of Waters and Associated Materials*, Environment Agency, 2002.
22 L. Fewtrell, D. Butler, F. Ali Memon, R. Ashley, A. Saul, in *Health Impact Assessment for Sustainable Water Management*, eds. L. Fewtrell and D. Kay, IWA Publishing, 2008, ch. 2, p. 29.
23 J.F. Shirlaw and S.G. Iyer, *Indian J. Vet. Sci.*, 1937, **7**, 231.
24 R.W.A. Girdwood, C.R. Fricker, D. Munro, C.B Shedden and P. Monaghan, *J. Hygiene*, 1985, **95**, 229.
25 S.E. Craven, N.J. Stern, E. Line, J.S. Bailey, N.A. Cox and P. Fedorka-Cray, *Avian Dis.*, 2000, **44**, 715.
26 J. Waldenström, T. Broman, I. Carlsson, D. Hasselquist, R.P. Acherberg, J.A.
27 P.S. Mead, L. Slutsker, V. Dietz, L.F. McCaig, J.S. Bresee, C. Shapiro, P.M. Griffen and R.V. Tauxe, *Emerging Infect. Dis.*, 1999, **5**, 607.
28 G.J. Medema, P.F.M. Teunis, A.H. Havelaar and C.N. Haas, *Int. J. Food Microbiol.*, 1996, **30**, 101.
29 WHO *Guidelines for drinking-water quality, third edition, Volume 1, Recommendations*, World Health Organization, Geneva, 2004.
30 J. Barker and M.V. Jones, *J. Appl. Microbiol.* 2005, **99**, 339.
31 MTP, *WC design and efficiency – briefing note relating to policy scenario objectives in policy brief*, Market Transformation Programme, 2006.
32 AAP, *Age 3 to 5: Beyond Toilet Training*, American Academy of Pediatrics, 2000.
33 G.K. Adak, S.M. Long and S.J. O'Brien, *Gut*, 2002, **51**, 832.
34 WHO *Guidelines for safe recreational water environments, Volume 1, Coastal and fresh waters*, World Health Organization, Geneva, 2003.

SUBJECT INDEX